THE CLOCKWORK UNIVERSE

THE
CLOCKWORK
UNIVERSE

*Isaac Newton, the Royal Society,
and the Birth of the Modern World*

EDWARD DOLNICK

HARPER

An Imprint of HarperCollins*Publishers*

www.harpercollins.com

HarperCollins books may be purchased for educational, business, or sales
promotional use. For information, please write: Special Markets Department,
HarperCollins Publishers, 10 East 53rd Street, New York, NY 10022.

FIRST EDITION

Designed by Eric Butler

Library of Congress Cataloging-in-Publication Data has been applied for.

ISBN: 978-0-06-171951-6

11   12   13   14   15   ov/rrd   10   9   8   7   6   5   4   3   2   1

For Lynn

The universe is but a watch on a larger scale.

—BERNARD DE FONTENELLE, 1686

CONTENTS

CHRONOLOGY

1543	Copernicus publishes *On the Revolutions of the Celestial Spheres*, which says that the planets circle the sun rather than the Earth
1564	Shakespeare born
1564	Galileo born
1571	Kepler born
1600	Shakespeare writes *Hamlet*
1609	Kepler publishes his first two laws, about the paths of planets as they orbit the sun
1610	Galileo turns a telescope to the heavens
1616	Shakespeare dies
1618–1648	Thirty Years' War
1619	Kepler publishes his third law, which tells how the planets' orbits relate to one another
1630	Kepler dies
1633	Inquisition puts Galileo on trial
1637	Descartes declares "I think, therefore I am," and, in the same book, unveils coordinate geometry
1642–1651	English Civil War
1642	Galileo dies

PREFACE

Few ages could have seemed less likely than the late 1600s to start men dreaming of a world of perfect order. Historians would later talk of the "Age of Genius," but the "Age of Tumult" would have been just as fitting. In the tail end of Shakespeare's century, the natural and the supernatural still twined around one another. Disease was a punishment ordained by God. Astronomy had not yet broken free from astrology, and the sky was filled with omens.

The only man-made light came from flickering flames and sputtering lanterns. Unless the moon was out, nights were dark and dangerous. Thieves and muggers prowled the streets—the first police forces lay far in the future—and brave souls who ventured outdoors carried their own lanterns or hired a "linkboy" to light the way with a torch made from a hunk of fat-soaked rope. The murder rate was five times as high as it is today.

Even in midday, cities were murky and grimy. Coal smoke left a "sooty Crust or Furr" on all it touched. London was one of the world's great cities and a center of the new learning, but it was, in one historian's words, "a stinking, muddy, filth-bespattered metropolis." Huge piles of human waste blocked city streets, and butchers added heaps of the "soyle and filth of their Slaughter houses" to the towering mounds.

Ignorance made matters worse. The same barges that brought vegetables to the city from farms in the countryside returned laden with human sewage, to fertilize the fields. When Shakespeare and his fellow investors built the Globe Theatre in 1599, the splendid new building held at least two thousand people but was constructed without a single toilet. Well over a century later, hygiene had scarcely improved. At about the time of Louis XIV's death in 1715, a new rule was put in place requiring that the corridors in the palace at Versailles be cleaned of feces once a week.

No one bathed, from kings to peasants. The poor had no choice, and the wealthy had no desire.* (Doctors explained that water opened the pores to infection and plague. A coat of grease and grime sealed disease away.) Worms, fleas, lice, and bedbugs were near-universal afflictions. Science would soon revolutionize the world, but the minds that made the modern world were yoked to itchy, smelly, dirty bodies.

On the public stage, all was crisis and calamity. Through the early part of the century, Germany had suffered through what would later be called the Thirty Years' War. The blandness of the name obscures the horror of a religious war where one raping, looting, marauding army gave way to another, endlessly, and where famine and disease followed close on the armies' heels. England had been convulsed by a civil war. In London in 1649, a shocked crowd looked on as the royal executioner lifted his axe high and chopped off the king's head. In the 1650s plague swept across Europe. In 1665 it jumped the Channel to England.

In the wings, the events that would reshape the world went on unnoticed. Few knew, and fewer cared, about a handful of

---

* The historian Jules Michelet described the Middle Ages as "a thousand years without a bath."

curious men studying the heavens and scribbling equations in their notebooks.

Humans had recognized nature's broad patterns from the beginning—night follows day, the moon waxes and wanes, the stars form their familiar constellations, the seasons recur. But they had noticed, too, that no two days were identical. "Men expected the sun to rise," wrote Alfred North Whitehead, "but the wind bloweth where it listeth." If people referred to "laws of nature," they had in mind not true laws but something akin to rules of thumb, guidelines subject to exceptions and interpretation.

Then, at some point in the 1600s, a new idea came into the world. The notion was that the natural world not only follows rough-and-ready patterns but also exact, formal, mathematical laws. Though it looked haphazard and sometimes chaotic, the universe was in fact an intricate and perfectly regulated clockwork.

From the cosmically vast to the infinitesimally small, every aspect of the universe had been meticulously arranged. God had created the world and designed its every feature, and He continued to supervise it with minute care. He had set the planets in orbit and lavished care on every one of a housefly's thousand eyes. He had chosen the perfect rate for the Earth's spin and the ideal thickness for a walnut's shell.

Nature's laws were vast in range but few in number; God's operating manual filled only a line or two. When Isaac Newton learned how gravity works, for instance, he announced not merely a discovery but a "universal law" that embraced every object in creation. The same law regulated the moon in its orbit round the Earth, an arrow arcing against the sky, and an apple falling from a tree, and it described their motions not only in general terms but precisely and quantitatively. God was a mathematician, seventeenth-century scientists firmly believed. He had written His laws in a mathematical code. Their task was to find the key.

My focus is largely on the climax of the story, especially Newton's unveiling, in 1687, of his theory of gravitation. But Newton's astonishing achievement built on the work of such titans as Descartes, Galileo, and Kepler, who themselves had deciphered paragraphs and even whole pages of God's cosmic code. We will examine their breakthroughs and false trails, too.

All these thinkers had two traits in common. They were geniuses, and they had utter faith that the universe had been designed on impeccable mathematical lines. What follows is the story of a group of scientists who set out to read God's mind.

Part One

CHAOS

Chapter One

LONDON, 1660

A stranger to the city who happened to see the parade of eager, chattering men disappearing into Thomas Gresham's mansion might have found himself at a loss. Who were these gentlemen in their powdered wigs, knee breeches, and linen cravats? It was too early in the day for a concert or a party, and this was hardly the setting for a bull-baiting or a prizefight.

With its shouting coachmen, reeking dunghills, and grit-choked air, London assaulted every sense, but these mysterious men seemed not to notice. Locals, then, for the giant metropolis left newcomers reeling. The men at Gresham's looked a bit like a theater crowd—and with the Puritans out of power and Oliver Cromwell's head on a pole in front of Westminster Hall, theaters *had* opened their doors again. But in that case where were the women? Perhaps the imposing building on the fashionable street concealed a gentlemen's gambling club? A high-class brothel?

Even a peek through a coal-grimed window might not have helped much. Amid the bustle, one man seemed to be spilling powder onto the tabletop and arranging it into a pattern. The man standing next to him held something between his fingers, small and dark and twitching.

The world would eventually learn the identity of these mysterious men. They called themselves natural philosophers, and they had banded together to sort out the workings of everything from pigeons to planets. They shared little but curiosity. At the center of the group stood tall, skeletally thin Robert Boyle, an aristocrat whose father was one of Britain's richest men. Boyle maintained *three* splendid private laboratories, one at each of his homes. Mild-mannered and unworldly, Boyle spent his days contemplating the mysteries of nature, the glories of God, and home remedies for an endless list of real and imaginary ills.

If Boyle was around, Robert Hooke was sure to be nearby. Hooke was hunched and fidgety—"low of stature and always very pale"—but he was tireless and brilliant, and he could build anything. For the past five years he had worked as Boyle's assistant, cobbling together equipment and designing experiments. Hooke was as bad-tempered and sharp-tongued as Boyle was genial. To propose an idea was to hear that Hooke had thought of it first; to challenge his claim was to make a lifelong enemy. But few questioned the magic in his hands. Hooke's latest coup was a glass vessel that could be pumped empty of air. What would happen if you put a candle inside? a mouse? a man?

The small, birdlike man was Hooke's closest friend, the ludicrously versatile Christopher Wren. Ideas tumbled from him like coins from a conjuror's fingertips. Posterity would know Wren as the most celebrated architect in English history, but he was renowned as an astronomer and a mathematician before he sketched his first building. Everything came easily to this charmed and charming creature. Early on an admirer proclaimed Wren a "miracle of youth," and he would live to ninety-one and scarcely pause for breath along the way. Wren built telescopes, microscopes, and barometers; he tinkered with designs for submarines; he built a transparent beehive (to see what the bees were up to) and a writing gizmo for making copies, with two pens connected by a wooden arm; he built St. Paul's Cathedral.

The Royal Society of London for the Improvement of Natural Knowledge, the formal name of this grab-bag collection of geniuses, misfits, and eccentrics, was by most accounts the first official scientific organization in the world. In these early days almost any scientific question one might ask inspired blank stares or passionate debate—Why does fire burn? How do mountains rise? Why do rocks fall?

The men of the Royal Society were not the world's first scientists. Titans like Descartes, Kepler, and Galileo, among many others, had done monumental work long before. But to a great extent those pioneering figures had been lone geniuses. With the rise of the Royal Society—and allowing for the colossal exception of Isaac Newton—the story of early science would have more to do with collaboration than with solitary contemplation.

Newton did not attend the Society's earliest meetings, though he was destined one day to serve as its president (he would rule like a dictator). In 1660 he was only seventeen, an unhappy young man languishing on his mother's farm. Soon he would head off to begin his undergraduate career, at Cambridge, but even there he would draw scarcely any notice. In time he would become the first scientific celebrity, the Einstein of his day.

No one would ever know what to make of him. One of history's strangest figures, Newton was "the most fearful, cautious, and suspicious Temper that I ever knew," in the judgment of one contemporary. He would spend his life in secrecy and solitude and die, at eighty-four, a virgin. High-strung to the point of paranoia, he teetered always on the brink of madness. At least once he would fall over the brink.

In temperament Newton had little enough in common with the other men of the Royal Society. But all the early scientists shared a mental landscape. They all lived precariously between two worlds, the medieval one they had grown up in and a new one they had only glimpsed. These were brilliant, ambitious, confused, conflicted men. They believed in angels and alchemy and the devil, *and* they believed that the universe followed precise, mathematical laws.

In time they would fling open the gates to the modern world.

Chapter Two

SATAN'S CLAWS

Scientists in the 1600s had set out to find the eternal laws that govern the universe, but the world they lived in was marked by precariousness.* Death struck often, and at random. "Any cold might be the forerunner of a terminal fever," one historian remarks, "and the simplest cut could lead to a fatal infection." Children died in droves, but no one was safe. Even for the nobility, life expectancy was only about thirty. Adults in their twenties, thirties, and forties dropped dead out of the blue, leaving their families in desperation.

London was so disease-ridden that deaths outnumbered births; only the constant influx of newcomers disguised that melancholy fact. Medical knowledge was almost nonexistent, and doctors were more likely to harm their patients than to heal them. Those who fell ill could do little more than choose from a reeking cupboard of quack remedies. One treatment for gout called for "puppy boiled up with cucumber, rue and juniper." As

---

* For convenience I will use the word *scientist,* though the word only came into use in the 1800s. The seventeenth century had not settled on a convenient term for these investigators. Sometimes they were called "natural philosophers" or "virtuosos."

late as 1699 the Royal Society was still debating the health benefits from "cows piss drank to about a pint."

The main alternative was woeful resignation. "I have had the misfortune of losing my dear child Johney he died last week of a feaver," a woman named Sarah Smyter wrote in a letter in 1717. "It tis a great trouble to me but these misfortunes we must submit to."

The mighty had no better options than the lowly. Many times they were worse off, because they were more likely to face a doctor's attentions. When Charles II suffered a stroke in 1685, his doctors "tortured him," one historian later wrote, "like an Indian at a stake." First the royal physicians drained the king of two cups of blood. Next they administered an enema, a purgative, and a dose of sneezing powder. They drained another cup of blood, still to no effect. They rubbed an ointment of pigeon dung and powdered pearls onto the royal feet. They seared the king's shaved skull and bare feet with red-hot irons. Nothing helped, and the king fell into convulsions. Doctors prepared a potion whose principal ingredient was "forty drops of extract of human skull." After four days Charles died.

Two killers inspired more fear than any others. One was plague, the other fire. Both killed swiftly and in huge numbers, but in different manners. Plague leaped stealthily from victim to victim. Its mystery made for its horror. "For what is the cause that this pestilence is so greatly in one part of the land, and not another?" one panicky writer had asked during an earlier epidemic. "And in the same citie and towne why is it in one part, or in one house, and not in another? and in the same house, why is it upon one, and not upon all the rest?"

Dance of the Skeletons
(1493)

Fire had scarcely any mystery about it. It terrified precisely because it killed spectacularly, mercilessly, and in plain view. In crowded, cramped cities built of wood and lit by flame, it was all but inevitable that somewhere a hot coal would fall from a stove or a furnace, or a candle would tumble against a curtain or onto a pile of straw. Once escaped, even a small fire could blaze up into an inferno that sped along like a leaping, crackling tsunami. Its desperate victims raced for their lives down one twisting alley after another, fleeing round this corner and down that street, trying to outrun a pursuer that grew ever more powerful as the chase continued.

The dread that these ancient enemies inspired never died away, for everyone knew that no lull could be counted on to last. Nor did anyone think of fire and plague as natural calamities, the way we think of earthquakes and volcanoes. The seventeenth century was God-fearing in the most literal sense. Natural disasters were divine messages, warnings to sinful mankind to change its ways lest an angry and impatient God unleash still further rounds of punishment. Even today insurance claims refer to earthquakes and floods as "acts of God." In the 1600s and long beyond, our ancestors invoked the same phrase, but they spoke of God's mysterious will with fright and cowering awe.

* * *

In that harsh age religion focused far more on damnation than on consolation. For scientists and intellectuals pondering the course of the universe and for the common man as well, fear of God shaped every aspect of thought. To study the world was to ponder God's plan, and that was daunting work.

Today *damn* and *hell* are the mildest of oaths, suitable responses to a stubbed toe or a spilled drink. For our forebears, the prospect of being damned to hell was vivid and horrifying. "People lived in continual terror of what they were told awaited them after death," wrote the historian Morris Kline. "Priests and ministers affirmed that nearly everyone went to hell after death, and described in greatest detail the hideous, unbearable tortures that awaited the eternally damned. Boiling brimstone and intense flames burned victims who, nevertheless, were not consumed but continued to suffer these unabating tortures. God was presented not as the savior but as the scourge of mankind, the power who had fashioned hell and the tortures herein and who consigned people to it, confining His affection to only a small section of His flock. Christians were urged to spend their time meditating upon eternal damnation in order to prepare themselves for life after death."

God, who knew all the details of how the future would unroll, had decided already who would be saved and who punished. He would not be bartered with. Whether a person led a good life or a depraved one would do nothing to alter God's verdict; to say otherwise would imply that lowly man could direct all-powerful God.

A book called *The Day of Doom* appeared in 1662, the same year the Royal Society received its formal charter, and explained such doctrines in verse. A huge success (it became the first bestseller in America), it dealt curtly with such matters as infants condemned to the flames of hell:

> *But get away without delay*
> *Christ pities not your cry:*
> *Depart to Hell, there may you yell,*
> *And roar eternally.*

Children learned these poems by heart. Eventually such views would prove too grim to prevail, but they lasted well into the 1700s. Jonathan Edwards lambasted New England congregations with his most famous sermon, "Sinners in the Hands of an Angry God," as late as 1741. "The God that holds you over the pit of hell, much as one holds a spider, or some loathsome insect over the fire, abhors you, and is dreadfully provoked: his wrath towards you burns like fire; he looks upon you as worthy of nothing else but to be cast into the fire; he is of purer eyes than to bear to have you in his sight; you are ten thousand times more abominable in his eyes than the most hateful venomous serpent is in ours."

This was standard doctrine. Worship of God began with acknowledging His might in contrast with human puniness. "Those are my best days when I shake with fear," John Donne declared. Few sins were too small to bring down God's wrath and to stir up soul-wrenching guilt. At age nineteen, in that same Royal Society year of 1662, Isaac Newton compiled a list of the sins he had committed in his life thus far. The tally, supposedly complete, listed fifty-eight items. Thoughts and acts were jumbled together, the one as bad as the other. One or two entries catch the eye—"Threatening my father and mother Smith [i.e., Newton's mother and stepfather] to burne them and the house over them"—but nearly all the list is mundane. "Making a mousetrap on Thy day." "Punching my sister." "Using Wilford's towel to spare my owne." "Having uncleane thoughts words and actions and dreamese." "Making pies on Sunday night." These sins may strike us as minor and commonplace. In Newton's eyes, they were deeply shameful betrayals of himself and his God.

In this self-lacerating respect, at least, Newton was far from unique. The writer and theologian Isaac Watts, who would grow up to compose such hymns as "Joy to the World," first revealed his talent in an acrostic he composed as a young boy, in the late seventeenth century. It began:

> **I** *am a vile polluted lump of earth*
> **S**o *I've continued ever since my birth,*
> **A**lthough *Jehovah grace does daily give me,*
> **A**s *sure this monster Satan will deceive me,*
> **C**ome *therefore, Lord, from Satan's claws relieve me.*

A second verse spelled out similar thoughts for the name Watts.

Chapter Three

THE END OF THE WORLD

In the 1650s and '60s the long-simmering fear of God's wrath grew acute. Every Christian knew his Bible, and everyone knew that the Bible talked of a day of judgment. The question was not whether the world would end but how soon the end would come. The answer, it seemed, was *very* soon.

Almost no one believed in the idea of progress. (The very scientists whose discoveries would create the modern world did not believe in it.) On the contrary, the nearly universal belief was that the world had been falling apart since Adam and Eve were banished from Eden. Now, it seemed, the fall had accelerated. From high and low, in learned sermons and shrieking pamphlets, men pointed out the signs that the apocalypse was near.

At some moment, at *any* moment, in one historian's summary, "The trumpet would sound, motion would cease, the moon turn to blood, the stars fall like withered leaves, and the earth would burn to the accompaniment of horrible thunders and lightnings." In the midst of this chaos, the dead would rise, and saint and sinner alike would receive a sentence that permitted no appeal and no pardon. In the minds of our ancestors, this was not rhetoric but fact. God had ordained it, and it would be so.

The debate about the timing of the end was intense and widespread. Today warnings that "the end is nigh!" are the stuff of television preachers and *New Yorker* cartoons. In the seventeenth century this was urgent business. Deciphering biblical prophecies was as much a mainstream, high-stakes concern then as poring over stock market figures is now. Similar waves of fear had arisen before, for no clear reason, and then died down just as mysteriously. That was no consolation. "Books on the Second Coming were written by the score during this period," one eminent historian observes, "and members of the Royal Society were preoccupied with dating the event." They proceeded methodically, looking for hidden meanings in biblical texts or manipulating numbers cited in one sacred passage or another.*

Many scholars and scientists pointed with alarm to a particular figure—1,260 years—that popped up at several different places in the Bible.† At some point in the past, they believed, the clock had started ticking. Twelve hundred and sixty years from that moment, the world would end. The question that obsessed the most powerful minds of the Royal Society was, when had the countdown begun? One frequently cited date—400 A.D., a time of "great apostasy" when true Christianity had been subverted. It did not demand the mathematical talent of Isaac Newton to see that 1,260 years from 400 A.D. brought one to the year 1660.

---

* Christopher Wren's father, a prominent cleric who also had a deep interest in mathematics, calculated the apocalypse in a different way. A list of the Roman numerals, in order from biggest to smallest—MDCLXVI— corresponded to the date 1666, which "may bode some ominous Matter, and perhaps the last End."

† They cited passages such as Revelation 11:3: "I will give power unto my two witnesses, and they shall prophesy a thousand two hundred and threescore days, clothed in sackcloth." Scholars took each day to represent a year.

Jesus himself had talked of the signs that would announce the final days. At the Mount of Olives the disciples had asked him, "What shall be the sign of thy coming, and of the end of the world?"

War and misery on Earth, Jesus had replied, and chaos in the heavens. "Nation shall rise against nation, and kingdom against kingdom: and there shall be famines, and pestilences, and earthquakes." And then, after still more afflictions, "Shall the sun be darkened, and the moon shall not give her light, and the stars shall fall from heaven."

Now, both on Earth and in the heavens, danger signs abounded. Adulterers, blasphemers, and disbelievers had transformed London into a modern-day Babylon. Such carryings-on were nearly inevitable, for a long, dour Puritan interlude had only recently ended. Following Charles I's execution in 1649, theaters had been closed, celebrations of Christmas banned, dancing at weddings outlawed.

After the restoration of the monarchy, in 1660, the mood changed utterly, at court and throughout the nation. Charles I had been earnest, stubborn, hidebound. Charles II was witty and restless, always ready to play another set of tennis, gamble on another hand of cards, chase after yet another beauty. Life at court was notoriously indulgent, with everyone from the king on down "engaged in an endless game of sexual musical chairs." (The "Merrie Monarch" kept mistresses by the score, but in these early years of his reign he observed a sort of fidelity, restricting himself to one mistress at a time.) For the wealthy and the well connected generally, the tone of the era was one of cynicism and self-indulgence.

Inevitably, many people ignored the prophets of doom or scoffed at their warnings and lamentations. But like distant sounds of shouting at a party, the signals of something amiss tainted the festive mood. God would not be mocked. In 1662, terrified onlookers in the English countryside reported that several women had given birth to monstrously deformed babies. At the same time a brilliant star had mysteriously appeared in the night sky. From Buckinghamshire, in southern England, came reports that blood had rained from the sky. The heavens were all askew, just as Jesus had warned.

Then, in the fall of 1664, Europe and England saw a comet ablaze in the heavens. To the seventeenth-century mind, this was bad news. (The word *disaster* comes from *dis*, as in *disgrace* or *disfavor*, and *astrum*, Latin for *star* or *comet*.) The sky, unlike the Earth, was a domain of order and harmony. Comets were ominous intruders, and they had been feared for millennia. "So horrible was it, so terrible, so great a fright did it engender in the populace," one eyewitness had written, about a comet in 1528, "that some died of fear; others fell sick . . . this comet was the color of blood; at its extremity we saw the shape of an arm holding a great sword as if about to strike us down. At the end of the blade there were three stars. On either side of the rays of this comet were seen great numbers of axes, knives, bloody swords, amongst which were a great number of hideous human faces, with beards and hair all awry."

Comets were cosmic warnings, signs of God's displeasure akin to lightning bolts but longer lasting. "The thick smoke of human sins, rising every day, every hour, every moment . . . [grow] gradually so thick as to form a comet," explained one follower of Martin Luther, ". . . which at last is kindled by the hot and fiery anger of the Supreme Heavenly Judge."

Unsettlingly, comets hung overhead for days before they disappeared. Where they went, or why, no one knew. Night after night, all one could do was check the sky to see if the dreaded visitor had appeared again and guess at what calamity it might foretell.

This newest comet refused to disappear. The fearsome sightings of 1664 persisted into November and then into December. On December 17, King Charles II and Queen Catherine waited late into the night to witness the spectacle for themselves. The public mood grew ever darker. In January, word came of an apparition near the comet—"a Coffin," floating in the sky, "which causes great anxiety of thought amongst the people."

An astrologer and member of the Royal Society, John Gadbury, warned that "this comet portends pestiferous and horrible winds and tempests." Another astrologer foresaw "a MORTALITY which will bring MANY to their Graves."

In March 1665, a *second* comet appeared.

Closer to home, the natural world seemed just as unsettled. Rumors and omens started with worrisome sightings—clouds of flies swarmed inside houses; ants smothered the roads; frogs clogged the ditches—and grew ever more lurid. Like Egypt in ancient days, England had angered God. Even the well educated passed along the latest news in horrified whispers, as frightened and fascinated as the most superstitious countrymen. "A deformed monster" had been born in London, the Spanish ambassador reported, "horrible in shape and color. Part of him was fiery red and part of him yellow. On his chest was a human face. He had the legs of a bull, the feet of a man, the tail of a wolf, the breasts of a goat, the shoulders of a camel, a long body and in place of a head a kind of tumor with the ears of a horse. Such monstrous prodigies are permitted by God to appear to mankind as harbingers of calamities."

The greatest scientists of the age, Isaac Newton chief among them, believed as fervently as everyone else that they lived in the shadow of the apocalypse. Every era lives with contradictions that it manages to ignore. The Greeks talked of justice and kept slaves. The Crusaders preached the gospel of the Prince of Peace and rode off to annihilate the infidels. The seventeenth century believed in a universe that ran like clockwork, entirely in accord with natural law, and also in a God who reached down into the world to perform miracles and punish sinners.

Many of the early scientists tended not to pay much heed to monsters and bloody rains, but they pored over their Bibles in an urgent quest to determine how much time remained. Robert Boyle, renowned today as the father of chemistry, studied the Bible not only in English but in Greek, Hebrew, and Chaldean, to ferret out hidden meanings. Newton himself owned some thirty Bibles in various translations and languages that he endlessly perused and compared one against another.

Every word in the Bible was meaningful, just as every twig and sparrow in the natural world offered up a clue to God's intent. The Bible was not a literary work to be interpreted according to one's taste, but a cipher with a single meaning that could be decoded by a meticulous and brilliant analyst. Newton devoted thousands of hours—as much time as he spent on the secrets of gravity or light—in looking for concealed messages in the dimensions of the Temple of Solomon and trying to match the prophecies in Revelation with the battles and revolutions of later days. "The fourth beast [in the book of Revelation] . . . was exceeding dreadful and terrible, and had great iron teeth, and devoured and brake in pieces, and stamped the residue with its feet," wrote Newton, "and such was the Roman empire."

* * *

With nearly everyone in agreement that the end had drawn near, the debate turned to just *how* the end would come. One faction maintained that the world would drown in a global flood, as it had in Noah's day; others held out for an all-consuming fire. The tide of fear rose ever higher as the ominous year 1666 appeared, because of the satanic associations of the number 666. Fear turned to panic when plague swooped down on England in 1665, a year ahead of schedule, and death carts began spilling their cargo into mass graves.

Chapter Four

"WHEN SPOTTED DEATH RAN ARM'D THROUGH EVERY STREET"

For a thousand years, whenever God lost patience with his creation, plague had swept across Europe. For a few hundred years, those waves of disease had taken on a fearsome rhythm, appearing and vanishing at intervals of roughly ten or twenty years. In the deadliest assault, from 1347 through 1350, plague killed twenty million people. Somewhere between one-third and one-half of all Europeans died in that three-year span.

England's population crashed so far that it did not return to its pre-plague level for four centuries. In Florence the dead lay piled in pits "like cheese between layers of lasagna," in the words of one repelled, stunned observer. The survivors could do little more than gape at the devastation. "Oh happy posterity," wrote the Italian poet Petrarch, "who will not experience such abysmal woe and will look upon our testimony as a fable."

This was the bubonic plague, a disease spread to humans by fleas that had bitten infected rats, though no one would know that for centuries. Plague sputtered along between full-fledged outbursts, claiming a few lives almost every year but seldom flaring out of control. For decades in the mid-1600s England had been granted a respite. Plague had devastated one European city or another through those years, but since 1625 it had spared London.

No city lay beyond reach, though, for plague traveled with ships, armies, and merchants—with any travelers who unknowingly brought rats and fleas with them. England had begun to grow rich in the seventeenth century, and much of its wealth was based on trade. From all over the world, ships brought tea and coffee, silk and china, tobacco and sugar, to England's teeming ports. Europe, in the meantime, had spent the 1650s and '60s watching helplessly as plague moved across the continent. Italy and Spain had succumbed first, then Germany. In 1663 and 1664, plague devastated Holland.

In England, all was quiet—a single plague death in London at Christmas, 1664; another in February; two in April. On April 30, 1665, Samuel Pepys mentioned plague in his diary for the first time.* Pepys was still young, just past thirty and newly embarked on a career as a Royal Navy bureaucrat. The diary that would one day become a world treasure was only a private diversion. Pepys's first reference to plague was brief, an afterthought following a cheery description of dinner and the state of his finances. He had gone through his account books and found, "with great joy," that he was richer than he had ever been in his life. Then a quick observation: "Great fears of the sicknesse here in the City, it being said that two or three houses are already shut up. God preserve us all."

It is hard to read that first, ominous passage without hearing

---

* *Pepys* is pronounced "peeps."

a horror movie's minor chords in the background. In the face of the calamity that lay ahead, Pepys's mention of "two or three" tragedies would come to sound almost quaint.

Plague killed arbitrarily, agonizingly, and quickly. "A nimble executioner," in the words of one frightened observer, it could kill a healthy man overnight. No one knew the cause; no one knew a cure. All that was known was that plague somehow jumped from person to person. The sick fell and died, and the not-yet-infected cowered and waited.

The first symptom could be as innocuous as a sneeze (the custom of saying "Bless you!" when someone sneezes dates from this era). Fever and vomiting followed close behind. Next came "the surest Signes," in the words of one pamphlet from England's epidemic of 1625, an onslaught of blisters on the skin and swellings beneath it. Blue or purplish spots about the size of a penny appeared first. Shortly after, angry red sores flared up, "as if one did burne a hole with a hot iron." Then followed the dreaded black swellings that marked the end. They bulged out from the neck, armpits, or groin, sometimes "no bigger than a Nutmeg . . . but some as bigge as a Man's fist." Victims oozed blood from the tender lumps and moaned in pain.

Once plague had struck, doctors could provide no help beyond a soothing word. Authorities focused all their attention on safeguarding the healthy. Those who had fallen ill were forbidden to step out of their homes; hired guards stood watch to keep the prisoners from escaping. Food was supposedly left on the doorstep by "plague nurses," but they were as likely to rob their dying charges as to help them.

Many houses where plague had struck were nailed shut, with those inside left to die or not as fate decreed. (Thus Pepys's reference to "houses already shut up.") Some slum tenements held half a dozen captive families. The houses of the condemned carried a large cross, marked on the door in red chalk, to warn others to keep away. Scrawled near the cross were the forlorn words, "Lord have mercy upon us."

On June 7, 1665, Pepys first saw "two or three such houses" for himself. On June 10, he decided it was time to write his will. On June 15, he noted that "the town grows very sickly . . . there dying this last week of the plague 112, from 43 the week before."

The numbers rose throughout the summer. Frightened Londoners discussed such patterns endlessly, as if they were trying to guess when a madman might strike next. On July 1 Pepys saw "seven or eight houses in Bazing-hall street shut up of the plague." On July 13 he recorded "above 700 dead of the plague this week."

The numbers were unreliable, for they were gathered by ignorant, despised old women called "searchers." Their twofold task was to count the dead and to seek out signs of plague among the living, so that officials could know which families to quarantine inside their homes. No one volunteered for such work. The searchers were poor women, on the dole, forced to take on their task by parish officials' threats to withhold their meager benefits. Shunned even in ordinary times, the searchers now bore the added stigma of carrying contagion with them. Passersby who saw the ragged women scurried to get away, and the law made sure that was easy. Searchers were required to carry a two-foot-long white wand as an emblem of office, and to walk close to the refuse channels in the street.

Shaky as the plague statistics were, the trend was unmistakable. Throughout the summer of 1665 the death toll rose from a few hundred a week in June to one thousand a week in July and then to six thousand a week by the end of August. London witnessed scenes that jarred even hardened witnesses. Children were more vulnerable than adults, but whole families fell ill in a matter of days. "Death was the sure midwife to all children, and infants passed immediately from the womb to the grave," wrote Nathaniel Hodges, a doctor who performed heroic service all through the plague time. "Some of the infected ran into the streets; while others lie half-dead and comatose, but never to be waked but by the last trumpet; some lie vomiting as if they had drunk poison; and others fell dead in the market."

At first, when the death rate was still low, Dr. Hodges had dared hope that the damage would stay in bounds. All such hopes were soon dashed. Plague was a "cruel enemy," Hodges lamented, like an army that "at first only scattered about its arrows, but at last covered the whole city with dead." Hodges told of priests in perfect health who went to comfort dying men and died alongside them. Doctors at the bedside keeled over next to their patients. Pepys heard about a now-commonplace disaster that had befallen an acquaintance. "Poor Will that used to sell us ale . . . , his wife and three children died, all, I think in a day."

Chapter Five

MELANCHOLY STREETS

The authorities flailed about in search of a solution. Plays, bull-baitings, and other entertainments were banned, because plague was known to be a disease of crowds. Was it the poor who carried this disease? The lord mayor tried to restrict the movements of the "Multitude of Rogues and wandering Beggars that swarm in every place about the City, being a great cause of the spreading of the Infection." Were animals the culprits? In the summer of 1665, the authorities called for the immediate killing of all cats and dogs. Orders went out to Londoners to kill "all their dogs of what sort or kind before Thursday next at ye furthest."* Thousands upon thousands of cats and dogs were killed. The result was to send the rat population soaring.

Nothing helped. Throughout the summer panicky crowds bent on escaping the contaminated city clogged the roads out of London. The poor stayed put. They had no money for travel and no place to go, but the rich and the merely well-to-do—doctors, lawyers, clergymen, and merchants—shoved their way into the

---

* The *ye* we are all familiar with ("Ye Fox and Hounds Tavern") was pronounced "the." The use of the letter *y* was a typographical convention, like ∫ for *s*.

scrum. Coaches and carriages knocked against one another, their horses pawing the mud, while heavy-laden wagons fought for position. The frenzied pack fighting through the narrow streets reminded one eyewitness of a terrified crowd in a burning theater. Some fled toward the Thames and tried to commandeer fishing boats, anything that could float and take them to safety. Those who managed to escape the city had to brave the residents of the countryside, who greeted the refugees with clubs and muskets.

The king and his brother, the Duke of York, fled London in early July. Most of the Royal Society had scattered by then, too, looking forward to a time "when we have purged our foul sins and this horrible evil will cease." Pepys sent his family away, but he himself retreated only as far as Greenwich. At the end of August he ventured on a long walk in the city. "Thus the month ends," he wrote, "with the plague everywhere through the Kingdom almost. Every day sadder and sadder news of its increase." In the last week of August, Pepys wrote, plague had claimed 6,102 lives in London alone.

Worse was to come.

September 1665 unnerved even Pepys. "Little noise heard day or night but tolling of bells," he lamented in a letter to a friend. (It was plague that had inspired John Donne to write, "Never send to know for whom the bell tolls; it tolls for thee.")

By now, with so many dead and so many gone, frenzy had given way to desolation. Grass grew in the streets of London. In place of the usual clamor of voices—street vendors had been banned, so newsboys and rat catchers and fish sellers no longer hawked their wares—silence reigned. "I have stayed in the city till above 7,400 died in one week, and of them above 6,000 of the plague," Pepys wrote, "and little noise heard day or night but tolling of bells; till I could walk Lombard Street and not meet twenty persons from one end to the other . . . ; till whole families, ten and twelve together, have been swept away."

Now there were too many dead for individual burials. At night death carts rattled along empty streets in search of bodies, the darkness penetrated only by flickering, yellow torchlights. Cries of "Bring out your dead!" echoed mournfully. But with death striking willy-nilly, there were too few men left to drive the carts, too few priests to pray over the victims, too few laborers to dig their graves. The carts made their way to mass burial pits and spilled in their cargo. Many Englishmen recalled the somber words of King Edward III, eyewitness to the horrific epidemic of an earlier day. "A just God now visits the sons of men and lashes the world."

And then, mysteriously and blessedly, it ended. In mid-October, Pepys reported six hundred fewer deaths than the week before. The survivors began the gloomy process of taking stock. "But Lord, how empty the streets are, and melancholy," wrote Pepys, "so many poor sick people in the streets, full of sores, and so many sad stories overheard as I walk, everybody talking of this man dead and that man sick, and so many in this place, and so many in that."

By the end of November 1665, people began to flock back to London. Within another month the epidemic had all but ended. The plague had claimed one-fifth of the city's population, a total of one hundred thousand lives.

Plague hit London harder than anywhere else, but all England had suffered. In some cases, as in the famous calamity in the village of Eyam, the cause could be pinpointed. In September 1665 a village resident named George Vicars opened a box. Someone in London had sent a gift. Vicars found a packet of used clothing, felt it was damp, and hung it before the fire to dry. The clothing was flea-infested. In two days Vicars was delirious, in four dead. The disease spread, but the local rector persuaded the villagers it would be futile to leave and dangerous to others besides. Outsiders left provisions at the village outskirts. The plague took a year to burn its way through Eyam. In the end, 267 of the village's 350 residents lay dead. (The rector who refused to flee, Reverend Mompesson, survived, but his wife did not.)

Nearly always, though, plague seemed to rise out of nowhere, like some ghostly poison. The university town of Cambridge, which had weathered several epidemics through the centuries, had a long-established policy in place. (Builders would one day unearth mass graves beneath the idyllic grounds.) When plague settled onto the town, the university shut down and sent its students and faculty away, to wait for a time when it would be safe to gather in groups again. In June 1665 plague struck Cambridge, and the university closed.

A young student named Isaac Newton gathered up his books and retreated to his mother's farm to think in solitude.

Chapter Six
FIRE

In the fateful year of 1666, a second calamity struck London. Perhaps God had *not* forgiven sinful mankind, after all. Perhaps those who had prophesied that the world would end in all-consuming fire had been right all along. Plague had been insidious and creeping; the new disaster was impossible to miss. But the Great Plague and the Great Fire had one similarity that outweighed the differences between them. Both were the work of an outraged God whose patience was plainly drawing to a close.

The fire burned out of control for four days, starting in the slums near London Bridge and quickly threatening great swaths of the city. One hundred thousand people were left homeless. Scores of churches burned to the ground. Iron bars in prison cells melted. The stunned survivors stumbled through the ruins of their smoldering capital and gazed in horror. Where a great city had stood just days before, one eyewitness lamented, "there is nothing to be seen but heaps of stones."

As for who had started the fire, everyone had a theory. Catholics had burned the city down, to weaken the Protestant hold on power. Foreigners had done it, out of envy and malice. The Dutch had done it, because Holland and England were at war, or the French had, because the French and the Dutch were allies.

The king himself even figured in the rumors—he was, people whispered, a monarch filled with hatred for London (which had clamored for his father's execution) and obsessed with building monuments to himself. What vengeance could compare with destroying the home of his enemies and then rebuilding it to suit his own taste?

But all such explanations were, in a sense, beside the point. To focus on who had set the fire was a mistake akin to confusing the symptoms of a disease with the illness itself. Any such calamity reflected the will of God. The proper question was not what tool God had seen fit to employ, but what had stirred his wrath. In any case, even the best of investigations would yield merely what Robert Boyle called "second causes." God remained the inscrutable "first cause" of everything. He had imposed laws on nature when he had created heaven and Earth, and ever afterward he had been free to change those laws or suspend them or to intervene in the world however he saw fit.

The fire began in the early hours of Sunday, September 2, 1666, in one of London's countless bakeshops. Thomas Farriner owned a bakery on Pudding Lane, deep in one of the mazes that made up London's crowded slums. He had a contract to supply ship's biscuits for the sailors fighting the Dutch. On Saturday night Farriner raked the coals in his ovens and went to bed. He woke to flames and smoke, his staircase afire.

Someone woke the lord mayor and told him that a blaze had started up near London Bridge. He made his way to the scene, reluctantly, and cast a disdainful eye at the puny flames. "Pish!" he said. "A woman might piss it out."

At that point, perhaps, the damage might still have been

confined. But a gust of wind carried sparks and flame beyond Pudding Lane to the Star Inn on Fish Street Hill, where a pile of straw and hay in the courtyard caught fire.

Everything conspired to create a disaster. For nearly a year London had been suffering through a drought. The wooden city was dry and poised to explode in flames, like kindling ready for the match. Tools to fight the blaze were almost nonexistent, and the warren of tiny, twisting streets made access for would-be firefighters nearly impossible in any case. (On his inspection tour the lord mayor found that he could not squeeze his coach into Pudding Lane.) Pumps to throw water on the flames were clumsy, weak contraptions, if they could be located in the first place and if someone could manage to connect them to a source of water. Instead firefighters formed lines and passed along buckets filled at the Thames. The contents of a leather bucket flung into an inferno vanished with a hiss and sizzle, like drops of water on a hot skillet.

Making matters harder still, London was not just built of wood but built in the most dangerous way possible. Rickety, slap-dash buildings leaned against one another like drunks clutching each other for support. On and on they twisted, an endless laby-rinth of shops, tenements, and taverns with barely a gap to slow the flames. Even on the opposite sides of an alleyway, gables tot-tered so near together that anyone could reach out and grab the hand of someone in the garret across the way. And since this was a city of warehouses and shops, it was a city booby-trapped with heaps of coal, vats of oil, stacks of timber and cloth, all poised to stoke the flames.

The only real way to fight the fire was to demolish the intact buildings in its path, in the hope of starving it of fuel. As the fire roared, the king himself pitched in to help with the demolition work, standing ankle-deep in mud and water, tearing at the walls with spade in hand. Slung over his shoulder was a pouch filled with gold guineas, prizes for the men working with him.

Propelled by strong winds, the fire roared along and then split in two. One stream of flames headed into the heart of the city, the other toward the Thames and the warehouses that lined it. The river-bound fire leaped onto London Bridge, in those days covered with shops and tall, wooden houses. At the water's edge, the flames reached heights of fifty feet. Panicky refugees stumbled through the mud and begged boatmen to carry them away.

On the fire's second night Pepys watched in shock from a barge on the Thames, smoke stinging his eyes, showers of sparks threatening to set his clothes afire. As he watched, the flames grew until they formed one continuous arch of fire that looked to be a mile long. "A horrid noise the flames made," Pepys wrote, and the crackling flames were only one note in a devil's chorus. People screamed in terror as they fled, blinded by smoke and ashes. House beams cracked like gunfire when they burned through. Hunks of roofs smashed to the ground with great, percussive thuds. Stones from church walls exploded, as if they had been flung into a furnace.

Through the next day things grew worse. "God grant mine eyes may never behold the like, who now saw above ten thousand houses all in one flame," wrote the diarist John Evelyn. "The noise and cracking and thunder of the impetuous flames, the shrieking of women and children, the hurry of people, the fall of towers, houses and churches, was like a hideous storm . . . near two miles in length and one in breadth. Thus I left it, burning, a resemblance of Sodom or the Last Day."

* * *

After four days, the wind finally weakened. For the first time, the demolition crews—who had resorted to blowing up houses with gunpowder—managed to corral the flames. As the fires burned down, Londoners surveyed the remnants of their city. Acre after acre was unrecognizable, the houses gone and even the pattern of roads and streets obliterated. People wandered in search of their homes, John Evelyn wrote, "like men in some dismal desert."

One Londoner hurried to St. Paul's Cathedral, long one of the city's landmarks but now only rubble. "The ground was so hot as almost to scorch my shoes," William Taswell wrote. The church walls had collapsed, and the bells and the metal areas of the roof had splashed onto the ground in molten puddles. Taswell loaded his pockets with scraps of bell metal as souvenirs.

Taswell was not the only visitor to St. Paul's. With their own homes destroyed, many Londoners had sought refuge in the huge, seemingly permanent cathedral. They found little but smoldering rocks. In desperate need of shelter, the refugees crawled inside the underground crypts and took their place alongside the dead.

The city itself lay silent and devastated. "Now nettles are growing, owls are screeching, thieves and cut-throats are lurking," one witness cried out. "And terrible hath the voice of the Lord been, which hath been crying, yea roaring in the City, by these dreadful judgments of the Plague and Fire which he hath brought upon us."

Chapter Seven

GOD AT HIS DRAWING TABLE

England's trembling citizens, it would eventually become clear, had the story exactly backward. The 1660s did not mark the end of time but the beginning of the modern age. We can hardly blame them for getting it wrong—the earliest scientists looked out at a world that was filthy and chaotic, a riot of noise, confusion, and sudden, arbitrary death. The sounds that filled their ears were a mix of pigs squealing on city streets, knives shrieking against grinders' sharpening stones, and street musicians sawing away at their fiddles. The smells were dried sweat and cattle, with a background note of sewage. Chronic pain was all but universal. Medicine was useless, or worse.

Who could contemplate that chaos and see order?

And yet Isaac Newton turned his attention to the heavens and described a cosmos as perfectly proportioned as a Greek temple. John Ray, the most eminent naturalist of the age, focused on the living world and saw just as harmonious a picture. Every plant and animal provided yet another example of nature's perfect design. Gottfried Leibniz, the German philosopher destined to become Newton's greatest rival, took the widest view of all and

reported the sunniest news. Leibniz took as his province Newton's stars and planets, Ray's insects and animals, and everything
in between. The great philosopher surveyed the universe in all
its variety and found, on every scale, an intricate, perfectly en-
gineered mechanism. God had fashioned the best of all possible
worlds.

One reason that seventeenth-century scientists had such faith
was mundane. Much of the mayhem all around them went un-
heeded, like the noise of screeching brakes and whooping sirens
on city streets today. But the crucial reasons ran deeper.

The founding fathers of science looked more or less like us,
under their wigs, but they lived in a mental world nothing like
ours. The point is not that they took for granted countless features of everyday life that we find horrifying or bewildering—
criminals should be tortured in the city square and their bodies
cut in pieces and mounted prominently around town, as a
warning to others; an excursion to Bedlam to view the luna-
tics made for ideal entertainment; soldiers captured in wartime
might spend the rest of their lives chained to a bench and rowing
a galley.

The crucial differences lay deeper than any such roster of
specifics can reveal. On even the broadest questions, our assumptions conflict with theirs. We honor Isaac Newton for his
colossal contributions to science, for example, but he himself regarded science as only one of his interests and probably not the
most important. The theory of gravity cut into the time he could
devote to deciphering hidden messages in the book of Daniel. To
Newton and all his contemporaries, that made perfect sense—
the heavens and the Earth were God's work, and the Bible was
as well, and so all contained His secrets. To moderns, it is as if

Shakespeare had given equal time to poetry and to penmanship, as if Michelangelo had put aside sculpture for basket weaving.

Look only at scientific questions, and the same gulf yawns. We take for granted, for instance, that we know more than our ancestors did, at least about technical matters. We may not have more insight into human nature than Homer, but unlike him we know that the moon is made of rock and pocked with craters. Newton and many of his peers, on the other hand, believed fervently that Pythagoras, Moses, Solomon, and other ancient sages had anticipated modern theories in every scientific and mathematical detail. Solomon and the others knew not only that the Earth orbited the sun, rather than vice versa, but they knew that the planets travel around the sun in elliptical orbits.

This picture of history was completely false, but Newton and many others had boundless faith in what they called "the wisdom of the ancients." (The belief fit neatly with the doctrine that the world was in decline.) Newton went so far as to insist that ancient thinkers knew all about gravity, too, including the specifics of the law of universal gravitation, the very law that all the world considered Newton's greatest discovery.

God had revealed those truths long ago, but they had been lost. The ancient Egyptians and Hebrews had rediscovered them. So had the Greeks, and, now, so had Newton. The great thinkers of past ages had expressed their discoveries in cryptic language, to hide them from the unworthy, but Newton had cracked the code.

So Newton believed. The notion is both surprising and poignant. Isaac Newton was not only the supreme genius of modern times but also a man so jealous and bad-tempered that he exploded in fury at anyone who dared question him. He refused to speak to his rivals; he deleted all references to them from his published works; he hurled abuse at them even after their deaths.

But here was Newton arguing vehemently that his boldest insights had all been known thousands of years before his birth.

The belief in ancient wisdom was overshadowed by other doctrines. By far the most important of the seventeenth century's bedrock beliefs was this: the universe had been arranged by an all-knowing, all-powerful creator. Every aspect of the world—why there is one sun and not two, why the ocean is salty, why lobster is delicious and deer are swift and gold is scarce, why one man died of plague but another survived—represented an explicit decision by God. We may not grasp the plan behind those decisions, we may see only disarray, but we can be certain that God ordained it all.

"All disorder," wrote Alexander Pope, was "harmony not understood." The world was an orderly text to those who knew how to read it, a tangle of blotches and squiggles to those who did not. God was the author of that text, and mankind's task was to study His creation, secure in the knowledge that every word and letter reflected divine purpose. "Things happen for a reason," we tell one another nowadays, by way of consolation after a tragedy, but for our forebears *everything* happened for a reason. At the core, the reason was always the same: God had willed it.* God was a daily presence and used events great and small—earthquakes, fires, victories in war, illness, a stumble

---

* God watched over the highest and the humblest. In Queen Elizabeth's reign the bishops of Canterbury, London, and Ely declared "this continued sterility in your Highness' person to be a token of God's displeasure towards us."

on the stairs—to demonstrate his wrath or his mercy. To imply that anything in the world happened by chance or accident was to malign Him. One should not speak of "fate," Oliver Cromwell had scolded, because it was "too paganish a word."

God saw every sparrow that falls, but that was only for starters. If God were to relax his guard even for a moment, the entire world would immediately collapse into chaos and anarchy. The very plants in the garden would rebel against their "cold, dull, inactive life," one Royal Society physician declared, and strive instead for "self motion" and "nobler actions."

To a degree we can scarcely imagine, the 1600s were a God-drenched era. "People rarely thought of themselves as 'having' or 'belonging to' a religion," notes the cultural historian Jacques Barzun, "just as today nobody has 'a physics'; there is only one and it is automatically taken to be the transcript of reality." Atheism was literally unthinkable. In modern times, we presume that either God exists or He doesn't. We can fight about the evidence, but the statement itself seems perfectly clear, no different in principle from *either there are mountains on the moon or there are not.*

In the seventeenth century no one reasoned that way. The idea that God might not exist made no sense. Even Blaise Pascal, one of the farthest-ranging thinkers who ever lived, declared flatly that it would be "absurd to affirm of an absolutely infinite and supremely perfect being" that He did not exist. The idea was meaningless. To raise the question would be to ponder an impossibility, like asking if today might come before yesterday.

For Newton and the other intellectuals of the day, God also had another aspect entirely. Not only had He created the universe and designed every last feature of every single object within it, not only did He continue to supervise His domain with an all-seeing, ever-vigilant eye. God was not merely a creator but a particular kind of creator. God was a mathematician.

That was new. The Greeks had exalted mathematical knowledge above all others, but their gods had other concerns. Zeus was too busy chasing Hera to sit down with compass and ruler. Greek thinkers valued mathematics so highly for aesthetic and philosophical reasons, not religious ones. The great virtue of mathematics was that its truths alone were certain and inevitable—in any conceivable universe, a straight line is the shortest distance between two points, and so on.* In the Greek way of thinking, all other facts stood on shakier ground. A mountain might be precisely 10,257 feet tall, but it could just as well have been a foot higher or lower. To the Greeks, historical facts seemed contingent, too. Darius was king of the Persians, but he might have drowned as a young boy and never come to the throne at all. Even the facts of science had an accidental feel. Sugar is sweet, but there seemed no particular reason it could not have tasted sour. Only the truths of mathematics seemed tamper-proof. Not even God could make a circle with corners.

Seventeenth-century thinkers rejected the Greeks' distinction between truths that have to be—*two and two make four*—and truths that happen to be—*gold is soft and easy to scratch.* Since every facet of the universe reflected a choice made by God, chance had no role in the universe. The world was rational and orderly. "It just so happens" was impossible.

---

* In 1823 a twenty-one-year-old Hungarian named Johann Bolyai conceived the inconceivable: a universe in which parallel lines meet and straight lines curve. In 1919 Einstein proved that we live in such a universe.

But the seventeenth century found its own reasons for regarding mathematics as the highest form of knowledge. The huge excitement among the new scientists was the discovery that the abstract mathematics that the Greeks had esteemed for its own sake turned out in fact to describe the physical world, both on Earth and in the heavens. On the face of it this was absurd. You might as well expect to hear that a newly discovered island had proved to be a perfect circle or a newfound mountain an exact pyramid.

Sometime around 300 B.C., Euclid and his fellow geometers had explored the different shapes you get if you slice a cone with a knife. Cut straight across and you get a circle; at an angle and you get an ellipse; parallel to one side, a parabola. Euclid had studied circles, ellipses, and parabolas because he found them beautiful, not useful. (In the Greek world, where manual labor was the domain of slaves, to label an idea "useful" would have been to sully it. Even to work as a tradesman or a shopkeeper was contemptible; Plato proposed that a free man who took such a job be subject to arrest.)

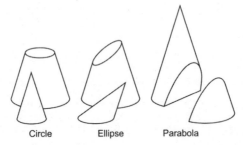

Circle   Ellipse   Parabola

Nineteen centuries later, Galileo found the laws that govern falling objects on Earth. After he showed the way, the discoveries came in a flood. Rocks thrown in the air and arrows shot from a bow travel in parabolas, and comets and planets move along ellipses exactly as if a colossal diagram from Euclid had been set among the stars. The universe had been meticulously arranged, Galileo and Kepler and Newton demonstrated, and the arrangement was the work of a brilliant geometer.

Then came an amazing leap. It was not simply that one aspect of nature or another followed mathematical law; mathematics governed *every* aspect of the cosmos, from a pencil falling off a table to a planet wandering among the stars. Galileo and the other seventeenth-century giants discovered a few golden threads and inferred the existence of a broad and gorgeous tapestry.

If God was a mathematician, it went without saying that He was the most skilled of all mathematicians. And since nature's laws are God's handiwork, they must necessarily be flawless—few in number, compact, elegant, and perfectly meshed with one another. "It is ye perfection of God's works that they are all done with ye greatest simplicity," Isaac Newton declared. "He is ye God of order and not of confusion."

The primary mission that seventeenth-century science set itself was to find His laws. The problem was that someone would first have to invent a new kind of mathematics.

Chapter Eight

THE IDEA THAT UNLOCKED THE WORLD

The Greeks had been brilliant mathematicians, but for centuries afterward that was the end of the story. Europe knew less mathematics in 1500, wrote Alfred North Whitehead, than Greece had in the time of Archimedes. A century later, matters had begun to improve. Descartes, Pascal, Fermat, and a small number of others had made genuine advances, though almost no one outside a tiny group of thinkers had any idea what they had been working on. The well educated in Newton's day knew Greek and Latin fluently, but a mathematical education typically ended with arithmetic, if it reached that far. "It was common," one historian writes, "for boys entering university to be unable to decipher the page and chapter numbers in a book." When Samuel Pepys took a high-level job as an administrator with the British navy, in 1662, he hired a tutor to teach him the mysteries of multiplication.

As able as the Greeks had been, they never found a way around one fundamental obstacle. They had nothing to say about motion. But if mathematics was going to describe the real world, it had to find a way to deal with moving objects. *If a bullet is shot into the air, how fast does it fly? How high does it rise?*

Alone on his mother's farm, twenty-three-year-old Isaac Newton set himself to unraveling the mystery of motion. (His mother hoped he would help run her farm, but he ignored her.) Newton's self-imposed task had two parts, and each was imposing. First, he had to invent a new language, some not-yet-known form of mathematics that would let him translate questions in English into numbers and equations and pictures. Second, he had to find a way to answer those questions.

It was a colossal challenge, but the Greeks' silence on the topic spoke more of distaste than of confusion. To the Greek way of thinking, the everyday world was a grimy, imperfect version of an ideal, unchanging, abstract one. Mathematics was the highest art because it was the discipline that, more than any other, dealt with eternal truths. In the world of mathematics, nothing dies or decays. The angles of a triangle add up to 180 degrees, and they did so a thousand years ago, and they will a thousand years in the future. To try to create a mathematics of change would be purposely to introduce impermanence and decline into the realm of perfect order.

Challenge a Greek mathematician with even the most elaborate question about a triangle or a circle or a sphere, then, and he would immediately have solved it. But triangles and spheres just sit there. Instead of drawing a picture of a sphere on a page, take a cannonball and shoot it into the sky. How high will it go? What path will it follow? How fast will it be moving when it crashes to the ground? In place of a cannonball, take a comet. If it passes overhead tonight, where will it be a month from now?

The Greeks had no idea. Until Isaac Newton and Gottfried

Leibniz came along to press the "on" button and set the static world in motion, no one else did, either. After they revealed their secrets, every scientist in the world suddenly held in his hands a magical machine. Pose a question that asked, *how far? how fast? how high?* and the machine spit out the answer.

The conceptual breakthrough was called calculus. It was the key that opened the way to the modern age, and it made possible countless advances throughout science. The word *calculus* conjures up little more, in the minds of most educated people today, than vague images of long equations and arcane symbols. But the world we live in is made of ideas and inventions as much as it is made of steel and concrete. Calculus is one of the most vital of those ideas. In an era that gave birth to the telescope and the microscope, to *Hamlet* and *Paradise Lost*, it was calculus that one distinguished historian proclaimed "by all odds the most truly revolutionary intellectual achievement of the seventeenth century."

Isaac Newton and Gottfried Leibniz invented calculus, independently, Newton on his mother's farm and Leibniz in the glittering Paris of Louis XIV. Neither man ever suspected for a moment that anyone else was on the same trail. Each knew he had made a stupendous find. Neither could bear to share the glory.

No hero ever rose from less auspicious roots than Isaac Newton. His father was a farmer who could not sign his name, his mother scarcely more learned. Newton's father died three months before his son was born. The baby was premature, so tiny and weak that no one expected him to survive; the mother was a widow, not yet thirty; the country was embroiled in civil war.

Newton did live, and lived to see honors heaped upon him. The fatherless boy, who was born on Christmas Day, believed throughout his life that he had been singled out by God. His story is so implausible that it almost seems that he might have been right. When Newton finally died, in 1727, at age eighty-four, a stunned Voltaire watched dukes and earls carry his casket. "I have seen a professor of mathematics, simply because he was great in his vocation, buried like a king who had been good to his subjects."

Newton's great opponent was a near contemporary—Leibniz was four years younger—and every bit as formidable as Newton himself. A boy wonder who had grown into an even more accomplished adult, Gottfried Leibniz had two strengths seldom found together: he was a scholar of such range that he seemed to have swallowed a library, and he was a creative thinker who poured forth ideas and inventions in half a dozen fields so new they had not yet been named. Even supremely able and ambitious men quailed at the thought of Leibniz's powers. "When one . . . compares one's own small talents with those of a Leibniz," wrote Denis Diderot, the philosopher/poet who had compiled an encyclopedia of all human knowledge, "one is tempted to throw away one's books and go die peacefully in the depths of some dark corner."

Leibniz was a lawyer and a diplomat by profession, but he seemed, almost literally, to know everything. He knew theology and philosophy and history, he published new theorems in mathematics and new theories in ethics, he taught himself Latin at seven and wrote learned essays on Aristotle at thirteen, he had invented a calculating machine that could multiply and divide (when rival machines could do no more than add and subtract). No subject fell outside his range. He knew more about China than any other European. Frederick the Great declared him "a whole academy in himself."

Leibniz's view of his own abilities was fully in line with Frederick's. On the rare occasions when praise was lacking, he supplied it himself. "I invariably took the first rank in all discussions and exercises, whether public or private," he remarked happily, recalling his school days. His favorite wedding gift to young brides was a collection of his own maxims. But somehow his vanity was so over-the-top, as was his flattery of the royal patrons he was forever wooing, that his exuberance seemed almost endearing. Throughout his long life, Leibniz retained the frantic eagerness of the smartest boy in fifth grade, desperately waving his hand for attention.

Newton and Leibniz never met. They would have made a curious-looking pair. Unlike Newton, who often slept in his clothes, Leibniz was a dandy who had a weakness for showy outfits with lace-trimmed cuffs, gleaming boots, and silk cravats. He favored a wig with long, black curls. Newton had a vain side, too, despite his austere manner—eventually he would pose for some seventeen portraits—and in his prime he cut a handsome figure. He was slim, with a cleft chin, a long, straight nose, and shoulder-length hair that turned silver-gray while he was still in his twenties. (Newton's early graying inspired his only recorded foray into the vicinity of humor. He had spent so much time working with mercury in his alchemical experiments, he once said, "as if from thence he took so soon that Colour.")

In appearance Leibniz was an odder duck. He was small, jumpy, and so nearsighted that his nose almost scraped the page as he wrote. Even so, he knew how to charm and chat, and he could set his earnestness aside. "It's so rare," the Duchess of Orléans declared happily, "for intellectuals to be smartly dressed, and not to smell, and to understand jokes."

Today we slap the word *genius* on every football coach who

Leibniz was greatly impressed by a demonstration of "a Machine for walking on water," which was apparently akin to this arrangement of inflatable pants and ankle paddles.

wins a Super Bowl, but both Newton and Leibniz commanded intellectual powers that dazzled even their enemies. If their talents were on a par, their styles were completely different. In his day-to-day life, as well as in his work, Leibniz was always riding off boldly in all directions at once. "To remain fixed in one place like a stake in the ground" was torture, he remarked, and he acknowledged that he "burned with the desire to win great fame in the sciences and to see the world."

Endlessly energetic and fascinated by everything under the sun, Leibniz was perpetually setting out to design a new sort of clock or write an account of Chinese philosophy and then dropping that project halfway through in order to build a better windmill or investigate a silver mine or explain the nature of free will or go to look at a man who was supposedly seven feet tall. At the same time that he was inventing calculus, in Paris in 1675, Leibniz interrupted his work and scurried off to the Seine to see an inventor who claimed he could walk on water.

No man ever had less of the flibbertigibbet about him than Isaac Newton. He had not a drop of Leibniz's impatience or wanderlust. Newton spent the eighty-four years of his life entirely within a triangle a bit more than one hundred miles on its longest side, formed by Cambridge, London, and Woolsthorpe, Lincolnshire, his birthplace. He made the short trip to Oxford for the first time at age seventy-seven, and he never ventured as far as the English Channel. The man who explained the tides never saw the sea.

Newton was a creature of serial obsessions, focusing single-mindedly on a problem until it finally gave way, however long that took. When an admirer asked him how he had come up with the theory of gravitation, he replied, simply and intimidatingly, "By thinking on it continually." So it was with alchemy or the properties of light or the book of Revelation. Week after week, for months at a stretch, Newton did without sleep and nearly without food ("his cat grew very fat on the food he left standing on his tray," one acquaintance noted).

"His peculiar gift was the power of holding continuously in his mind a purely mental problem until he had seen straight through it," wrote John Maynard Keynes, who was one of the first to examine Newton's unpublished papers. "I fancy his pre-eminence is due to his muscles of intuition being the strongest and most enduring with which a man has ever been gifted." An economist of towering reputation and intelligence, Keynes could only marvel at Newton's mental stamina. "Anyone who has ever attempted pure scientific or philosophical thought knows how one can hold a problem momentarily in one's mind and apply all one's powers of concentration to piercing through it, and how it will dissolve and escape and you find that what you are surveying is a blank. I believe that Newton could hold a problem in his mind for hours and days and weeks until it surrendered to him its secret."

Nothing diverted Newton. To test whether the shape of the

Newton's journal, showing his experiments on his own eye. Reproduced by kind permission of The Syndics of Cambridge University Library.

eyeball had anything to do with how we perceive color, Newton wedged a bodkin—essentially a blunt-ended nail file—under his own eyeball and pressed hard against his eye. "I took a bodkin & put it betwixt my eye & ye bone as neare to ye backside of my eye as I could," he wrote in his notebook, as if nothing could be more natural, "and pressing my eye with ye end of it . . . there appeared several darke and coloured circles." Relentlessly, he followed up his original experiment with one painful variation after another. What happened, he wondered, "when I continued to rub my eye with ye point of ye bodkin"? Did it make a difference "if I held my eye and ye bodkin still"?

In his zeal to learn about light, Newton risked permanent darkness.

Chapter Nine

EUCLID AND UNICORNS

In the early days of the Royal Society nearly anyone could attend its weekly meetings. Geniuses sat side by side with gentleman amateurs. The Society was less an ivory tower than a place to see and be seen. Giants like Robert Boyle and Christopher Wren presented their newest work, and so did such men as Sir Kenelm Digby, notable mainly for his faith in a potion called "weapon salve." Digby claimed he had used the salve to cure a man injured in a duel and given up for dead by the king's surgeons. The mysterious ointment included some unlikely ingredients— "moss from the skull of an unburied man," for one*—but the treatment was even odder than the medicine. Weapon salve was applied not to the wound but to the sword that had inflicted it, even if sword and victim were miles apart. (The wound itself was covered with a clean piece of linen and left alone, which in those pre-antibiotic days was probably a good thing.)

Along with stories of miraculous cures, tales of faraway lands were always popular. So was show-and-tell. On one October day in 1660, "the Society received a present of a living chameleon," after which Wren gave a talk about Saturn's rings. At another

---

* Digby assured his audience that "there is great quantity of it in Ireland."

meeting in 1660, the Society gravely scrutinized a unicorn's horn and then tested the ancient belief that a spider set down in the middle of a circle made from powdered unicorn's horn would not be able to escape. (The spider, unfazed, "immediately ran out several times repeated.")

Spiders turned up more often than one might have expected. On a winter afternoon in 1672 Isaac Newton made his first formal presentation to the Society. (Reclusive as always, Newton stayed away while someone else read aloud a paper he had sent.) Newton explained how he had found, using prisms, the true nature of light. White light was not pure but was made up of all the colors of the rainbow. This discovery was one of the milestones in the history of science. Newton's paper followed one on tarantula bites.

The Royal Society quickly accumulated so many strange and wondrous objects that it set up a museum. Visitors ogled such marvels of nature as "a tooth taken out of the womb of a woman, half an inch long," and "a piece of bone voided with his urine by Sir William Throgmorton."

Meetings were a hodgepodge because for every genius there was a crank or a charlatan. The surprise, from today's vantage point, is that so often the genius and the crank were the same person. Robert Boyle, for instance, was not merely a brilliant scientist and the most respected member of the Royal Society in its first decades but also the very model of circumspection and respectability. Boyle believed that the best cure for cataracts was to blow powdered, dried human excrement into the patient's eyes.

Ideas like that, and even more outlandish ones, were perfectly respectable. Three hundred years ago the boundary that separates the possible from the impossible was far fuzzier than it is today. In 1670 the Royal Society thrilled to reports of a new invention from Europe, a "flying chariot" that moved through the air powered by oars and a sail. Such optimism had its roots in genuine discoveries. Explorers had recently found entire "new" continents. The telescope had revealed astonishing new worlds, and the microscope, which was even newer, had shown that *this* world contained a multitude of unsuspected wonders. The humblest drop of pond water teemed with life.

The Royal Society's response to Kenelm Digby's claims for his magical "weapon salve" shows how much even learned men were prepared to believe. Since reliable men vouched for Digby's remedy, one highly regarded Society member remarked, "I need not be solicitous of the Cause." The world was so full of marvels, in other words, that the truly scientific approach was to reserve judgment about what was possible and what wasn't, and to observe and experiment instead. Digby's supposed cure strikes modern ears as a relic from an older, superstitious age. His contemporaries drew precisely the opposite moral—taking Digby's claims seriously displayed not backwardness and credulity but up-to-the-minute open-mindedness.

John Locke, a philosopher of decidedly levelheaded views (and, incidentally, a friend of Isaac Newton), considered it likely that the seas contained mermaids. Learned journals in the second half of the seventeenth century published articles with titles that sound like headlines from an ancient *National Enquirer.* "A Girl in Ireland, who has several Horns growing on her Body," "Description of an Extraordinary Mushroom," "Of Four Suns, which very lately appear'd in France."

Anything was possible.

We think of scientists as chucking out old ideas when something newer and more plausible comes along, but that is not the usual pattern. More often, scientists take up the new but cling to the old as well. In science's early days that was emphatically so. That made for some unlikely pairings. New ideas and old shared space in the same mind, like tattooed teens and hard-of-hearing dotards in uneasy coexistence in the same apartment.

Boyle, for example, held peculiar ideas about dead men and hangings. Eight times a year, on Hanging Days, immense crowds swarmed to London's gallows to see the show. The "Tyburn tree" had room for twenty-four swaying bodies at a time. Hanging Days were holidays, and crowds in dense, jolly packs lined the route from the prison gate to the gallows, like spectators at a parade. "All the Way, from Newgate to Tyburn, is one continued Fair, for Whores and Rogues of the meaner sort," one observer noted. The condemned man rode past the gawkers in a cart, seated atop his own coffin, his hands in manacles and his neck in a noose.

The throng at Tyburn might number twenty thousand. The rich, perched on wooden bleachers, had the best views. The poor fought for position. The crowd roared its approval for those prisoners who managed a defiant last word or a jaunty wave. A hangman who had been slipped a coin or two might make sure that his victim died quickly, but some condemned men twisted and choked, still half alive, at the end of a dangling rope. It made for extra excitement if the victim's friends flung themselves at his swinging body, frantically tugging his legs downward to try to speed his death.

For nearly everyone the spectacle itself was lure enough. Boyle and other connoisseurs knew better. One of the hangman's perks was the right to auction off souvenirs. Death ropes sold in one-foot lengths. But the hangman's most coveted trophies were an executed prisoner's severed hands, because a hand's "death sweat" held the power to heal. Robert Boyle, a giant of science, recommended this cure for those afflicted with a goiter.

The greatest figures in the pantheon of science gave equal weight to discoveries that we still celebrate and to ideas that strike us as mad. Take René Descartes, the brilliant mathematician and philosopher. He was one of the most important scientists of the generation just before Newton. If science is a cathedral, it was Descartes who set many of its foundation stones in place. Descartes was the ultimate skeptic, so reluctant to take anything for granted that he worried that the world and everything in it might simply be his dream. But he proposed a careful, scientific explanation for the well-known fact that if a person had been murdered and the killer later approached the victim's body, the corpse "identified" its killer by gushing blood.

William Harvey, renowned to this day for explaining how blood circulates in the body, was another who discovered the new while adhering to the old. Harvey was a contemporary of Descartes, which is to say that both men came of age at a time when the belief in witches was at its high point. Everyone knew a great deal about witches. They knew, for instance, that witches rubbed their bodies with "devil's grease," made from the fat of murdered babies, so that they could slither their way through tiny cracks into their victims' homes. They knew, as well, that witches had animal companions, cats or toads or rats, provided by Satan and magically able to do their mistress's evil bidding. Harvey, a man who straddled two ages, painstakingly dissected one witch's diabolical toad to see if he might find anything supernatural.

Alchemy, which was a scientific quest for a magical-sounding goal, provides perhaps the most striking example of the coexistence of old and new. The aim was to find a substance called the "philosopher's stone," despite its name a liquid, which held the power to transform ordinary substances into silver and gold and to convey immortality to anyone who drank it. A devout belief in alchemy was standard in the seventeenth century, but no one exceeded Isaac Newton in persistence. His small, crabbed handwriting fills notebook after notebook with the records of his alchemical experiments. In all Newton lavished some half million words on alchemy, about as many as in *War and Peace*.

He and countless other researchers spent long hours at their flasks and fires mixing potions according to closely guarded recipes. (Leibniz's only fear was that if gold became too readily available its price would fall.) An assistant watched Newton's experiments with reverence but without understanding. "Whatever his aim might be, I was not able to penetrate into, but his Pains, his Diligence at those Times made me think he aimed at something beyond the Reach of human Art & Industry."

A peek inside Newton's notebooks would have left an observer scarcely more enlightened. He never spoke of anything as crass as growing rich; his focus, it seems, was solely on uncovering nature's secrets. In any case, alchemical formulas were too valuable to state openly. All the language was encoded—"Saturn" stood for "lead," for instance—and the procedures sound like something from an X-rated Hogwarts spell-book. Newton jotted down recipes with such ingredients as "the Green Lion" and "the menstrual blood of the sordid whore."

The language is so strange, and Newton's scientific reputation is so high, that the temptation is to assume that the odd phrases merely indicate the difficulty of describing new techniques in an antique vocabulary. And it is true that in time alchemy gave rise to chemistry, and that Newton's approach to alchemy was methodical and absolutely rigorous. But it would be a mistake to conclude that Newton was a chemist in a sorcerer's hat.

On the contrary, Newton started out by studying chemistry but abandoned it in favor of what he saw as the deeper mysteries of alchemy. This was effectively a return to the past. Chemistry dealt with matter-of-fact questions like what salt is made from. Alchemy sought to explain the invisible forces of living nature. This was sacred, secret research. Throughout his long life Newton hardly breathed a word of what he was up to, and no wonder. "Just as the world was created from dark Chaos . . . ," he confided in a notebook, "so our work brings forth the beginning out of black chaos."

Newton's theological and alchemical writings went largely unexamined for two centuries after his death. In 1936, John Maynard Keynes purchased a trove of Newton's notes at auction. He read aghast. Newton was not the first inhabitant of the modern world, Keynes declared, but "the last of the Babylonians and Sumerians, the last great mind which looked out on the visible and intellectual world with the same eyes as those who began to build our intellectual inheritance rather less than ten thousand years ago."

Scientists tend to have little interest in history, even the history of their own subject. They turn to the past only to pluck out the discoveries and insights that turned out to be fruitful—Boyle, for instance, is known today for "Boyle's law," relating pressure and volume in gases—and they toss the rest aside.

In fields where the notion of progress is indisputable, such disdain for the past is common. The explanation is not so much anti-intellectualism as impatience. Why study ancient errors?

So scientists ignore most of their forebears or dismiss them as silly codgers. They make exceptions for a tiny number of geniuses whom they treat as time travelers from the present day, thinkers just like us who somehow found themselves decked out in powdered wigs.

But they were *not* like us.

Chapter Ten

THE BOYS' CLUB

Science today is a grand and formal enterprise, but the modern age of science began as a free-for-all. The idea was to see for yourself rather than to rely on anyone else's authority. The Royal Society's motto was "Nullius in Verba," Latin for, roughly, "Don't take anyone's word for it," and early investigators embraced that freedom with something akin to giddiness.

The meetings of the Royal Society in its young days sound like gatherings of a group of very smart, very reckless Cub Scouts. Society members gathered in a large room with a bare table and a roaring fire. In a group portrait, the men—the company was all male—would have looked more or less alike, but that was largely because everyone wore wigs. (In England and France, fashion followed the court. When Charles II began to go gray, and when the Sun King's hairs began to clog the royal hairbrush, the monarchs donned wigs, and soon no gentleman in Europe would venture out in public in his own hair.)

Half a dozen chairs, reserved for important visitors, sat empty on most days, while spectators jostled for space on two wooden benches. Seating was catch as catch can. New arrivals found places "as they think fit, and without any Ceremony," one French visitor wrote in amazement, "and if any one comes in after the

Society is fixed, no Body stirs, but he takes a Place presently where he can find it, so that no Interruption may be given to him that speaks." Whisperers were hushed indignantly.

The highlights, most weeks, were "demonstrations," the livelier the better. Hooke and Boyle carried out a long series of experiments to explore "the expansive forces of congelation"—they put water in a glass tube and froze it—and then everyone settled in to watch the tubes break "with a considerable noise and violence." Noise was always a great selling point. The members of the Royal Society were forever studying giant hailstones, for instance, in the hope that they would explode with a deafening crack when thrown into the fire. As a bonus, some hailstones had a strange shape or color. In those cases, the scientists' descriptions took on the tone of a "Ripley's Believe It or Not" item about a potato in the shape of a donkey.

Hooke had a particularly admired touch. He had figured out how to pump the air from a bell jar. (Official credit for building the air pump went to Boyle, for several years Hooke's employer.) Now he carried out experiment after experiment while his fellow scientists watched enthralled. "We put in a snake but could not kill it," one onlooker wrote perplexedly, but a chicken made a better show. "The chick died of convulsions outright, in a short space." What was the magical substance in ordinary air that living creatures needed in order to keep breathing, and why did some animals need more of it than others?

Soon Hooke and the others moved beyond experiments with birds and mice (and, less dramatically, with burning candles, which also seemed to need to "breathe"). On May 7, 1662, the Society needed something out of the ordinary for a particularly distinguished guest, Prince Rupert of the Rhine, cousin to the king. Out came the much-loved air pump. "We tried several experiments of Mr. Boyle's Vaccuum," wrote the diarist John Evelyn, who was in attendance. But what to put inside? Another mouse?

Robert Hooke had a better idea. "A man thrusting in his arm"—this was Hooke himself—"upon exhaustion of the air had his flesh immediately swelled, so as the blood was neere breaking the vaines, & unsufferable," Evelyn noted contentedly. "He drawing it out, we found it all speckled."

Transfusions made even better theater. On a November afternoon in 1667, forty witnesses crowded into the Society's meeting room to watch a blood transfusion from a sheep to a human. The subject was one Arthur Coga, "who, hearing that the Society were very desirous to try the experiment of transfusion upon a man, and being in want of money, offered himself for a guinea, which was immediately accepted on the part of the Society."

Coga had studied divinity at Cambridge but had suffered some kind of mental breakdown. That combination of credentials made Coga a perfect subject—his word could be trusted, since he was a gentleman, and he was mad, so he was intriguing. The hope was that the blood transfusion would cure him, though no one had any very good reason to think that might happen. While the crowd looked on, a surgeon made an incision into the sheep's leg and another into Coga's arm and then maneuvered a thin, silver pipe into place between them.

For two minutes blood passed from the sheep into Coga's body. Remarkably, Coga survived (although he did not recover his sanity). "After the operation the patient was well and merry," the surgeon reported, "and drank a glass or two of [wine] and took a pipe of tobacco in the presence of forty or more persons; then went home, and continued well all day."

* * *

Sheep to man blood transfusion.
Wellcome Library, London.

For the spectators who jostled one another for a better view of Arthur Coga's throbbing arm, every element of the scene before them was noteworthy. The experiment itself was new and untested, but the Royal Society's whole approach to the pursuit of knowledge constituted a much vaster, more important experiment.

Experiments were something new. The Society's devotion to this innovative way of probing nature amounted to a call for people to think for themselves. That idea, which seems like the merest common sense to us, struck onlookers at the time as dangerous and obviously misguided.

It's always the case that history is a tale told by the victors. But the triumph of the scientific worldview has been so complete that we've lost more than the losing side's version of history. We've lost the idea that a view different from ours is even possible. Today we take for granted that *originality* is a word of praise. *New* strikes us as nearly synonymous with *improved*. But for nearly all of human history, a new idea was a dangerous idea. When the first history of the Royal Society was written, in 1667, the author felt obliged to rebut the charge that "to be the Author of new things is a crime." By that standard, he argued, whoever raised the first house or plowed the first field could have been deemed guilty of introducing a novelty.

Most people would have agreed with the Spanish ruler Alfonso the Wise, who had once decreed that the only desirable things in this world were "old wood to burn, old wine to drink, old friends to converse with, and old books to read." The best way to learn the truth, it was often observed, was to see what the authorities of the past had decreed. This was the plainest common sense. To ignore such wisdom in favor of exploring on one's own was to seek disaster, akin to a foolish traveler's taking it in his head to fling the captain overboard and grab the ship's wheel himself.

Through long centuries the mission of Europe's great universities had been, in the words of the historian Daniel Boorstin, "not to discover the new but to transmit a heritage." (In the fourteenth century Oxford University had imposed a rule that "Bachelors and Masters of Arts who do not follow Aristotle's philosophy are subject to a fine of 5 shillings for each point of divergence.") The intellectual traits that we esteem today—like independence and skepticism—were precisely those traits that the Middle Ages feared and scorned.

That deference to authority had religious roots, as did nearly every aspect of medieval life. Good Christians showed their faith partly by their willingness to believe in the unbelievable. In a world riddled with miracles and mysteries, where angels and demons were as real as cats and dogs and where every illness and good harvest showed God's hand, skepticism was only a step from heresy. Who would set limits on the marvels the world contains? No one but an infidel.

So experiments had two linked drawbacks. To insist on

making one's own investigations was bad in itself, because it veered on impiety. In addition, looking for oneself meant second-guessing the value of eyewitness testimony. And for longer than anyone could remember, eyewitness testimony—whether it had to do with blood raining from the sky or the birth of half human/half animal monsters—had trumped all other forms of evidence. To accept such testimonials marked a person not as gullible or unsophisticated but as pious and thoughtful. To question such testimonials, on the other hand, the historians Lorraine Daston and Katharine Park remark, was "the hallmark of the narrow-minded and suspicious peasant, trapped in the bubble of his limited experience."

Augustine had laid out the argument many centuries before. "For God is certainly called Almighty for one reason only," he had written. That reason was perfectly plain: "He has the power to create so many things which would be reckoned obviously impossible" if not for the eyewitnesses who could swear to their truth.

The believers' task, then, was to defer to authority and refrain from asking questions, literally to "take it on faith." Augustine railed against the sin of curiosity with a fury and revulsion that, to modern ears, sound almost unhinged. Curiosity was, he wrote, a form of lust as despicable as any lusting of the flesh. The "lust to find out and know" was a perversion born of the same evil impulse that leads some people to peek at mutilated corpses or sneak into sideshows and stare at freaks. God intended that some mysteries remain beyond the bounds of human insight. Did not the Bible warn that "what the Lord keeps secret is no concern of yours; do not busy yourself with matters that are beyond you"?

Augustine's denunciation of curiosity prevailed for a thousand years. To seek to unravel nature's mysteries was to aspire to see the world with perfect clarity, and such insight was reserved for God alone. Pride was the great danger. "Knowledge puffeth up," Corinthians declared, and humankind had a duty to bear that rebuke constantly in mind. When the early scientists finally presumed to challenge that age-old dogma, traditionally minded thinkers sputtered in fury. No testimony was good enough for these maddening newcomers. "If the wisest men in the world tell them that they see it or know it; if the workers of miracles, Christ and his apostles, tell them that they see it; if God himself tells them that He sees it," one theologian thundered, "yet all this does not satisfy them unless they may see it themselves."

So the Royal Society's emphasis on experiments was a startling innovation. And experiments had still another feature that made them suspect. Experiments were by definition artificial. How could anyone draw universal, valid conclusions from special, manufactured circumstances? The problem with the new scientists' approach wasn't so much that they insisted on looking at nature rather than at books; the problem was that, not content with looking at the world, they insisted on manipulating it.

Premodern thinkers had studied the natural world closely. Astrologers scrutinized the night sky; botanists and doctors took notes on every plant that grew. But that had been a matter of observing and arranging rather than devising new questions to ask. The investigator's task had always been seen as akin to that of a librarian or a museum curator. For millennia, in one historian's words, an intellectual's "first duty" had been "absorbing, classifying, and preserving the known rather than exploring pastures new."

The new scientists, a less patient bunch, preferred the creed of their predecessor Francis Bacon, a contemporary of Shakespeare and the first great advocate of experimentation.* Nature must be "put to the torture," Bacon had declared. No doubt the image came quickly to mind in an age that coerced confessions by stretching prisoners on the rack or crushing their fingers in thumbscrews.

For the boisterous men of the Royal Society, spying on nature from behind a curtain was entirely too passive. Experiments had the great advantage that they let you *do* something. Preferably something dangerous. Hooke eventually managed to build a vacuum chamber so large that he could climb inside. Then, while the members of the Royal Society looked on with fascination, he gave the signal to pump the air out. The pump malfunctioned before Hooke could suffocate, but he did manage to render himself dizzy and temporarily deaf.

---

* Bacon's zeal for experimentation may have done him in. On a winter's day when he happened to be in the company of the royal physician, Bacon suddenly had the bright idea that perhaps snow could preserve meat. "They alighted out of the coach and went into a poor woman's house at the bottom of Highgate hill, and bought a fowl," wrote the memoirist John Aubrey, and Bacon stuffed the bird with snow. Bacon came down with what proved to be a fatal case of pneumonia. He blamed the snow but noted on his deathbed that the story had a bright side. "As for the experiment itself, it succeeded excellently well."

Chapter Eleven

TO THE BARRICADES!

The brilliant, frenetic Robert Hooke was a natural performer who took for granted that the best way to entertain an audience was to place himself in front of it. But the Royal Society experiments, which Hooke was charged with organizing, had a purpose beyond theatrics. The experiments also served as a call to arms against the old ways. The first rallying cry, as we have seen, was "Out of the library, into the laboratory." The second crucial message was "In plain sight." Ideas would be tested in the open, in front of witnesses. If an insight seemed genuine, other experimenters could test it for themselves.

This was an innovation. Until the mid-1600s everyone had always taken for granted that a person who made a discovery should keep the knowledge to himself, as secret as a treasure map, rather than give his fortune away by revealing it to the world. A plea from a mathematician named Girolamo Cardano, written about a century before the Royal Society's birth, highlighted the old attitude. Cardano wanted another mathematician to share a formula with him. "I swear to you by God's Holy Gospels and as a true man of honor, not only never to publish your discoveries, if you teach me them," Cardano begged, "but I also promise you, and I pledge my faith as a true Christian, to note them down in code, so that after my death no one will be able to understand them."*

The Royal Society pushed for a radically new approach: knowledge would advance more quickly if new findings were discussed openly and published for all to read. Thinkers would inspire one another, and ideas would breed and multiply. Robert Boyle made the most forceful argument against secrecy. A thinker who concealed his discoveries was worse than a miser who hoarded his gold, Boyle maintained, because the miser had no choice but to cling to his treasure. To give it away was to lose it. Thinkers had no such excuse, because ideas were not like gold but "like torches, that in the lighting of others do not waste themselves." With ideas as with flames, in fact, to share meant to *create* light.

Boyle insisted that this was ancient wisdom. "Our Saviour assureth us that it is more blessed to give than to receive," he reminded his fellow scientists, but this was a hard lesson to absorb. It remains hard today. To have found a secret that others are still scrabbling around for is to have a very special kind of private property. Modern physicists all know, and identify with, the story of Fritz Houtermans. In 1929 Houtermans wrote up a pioneering paper on fusion in the sun. The night he finished the work, he and his girlfriend went for a stroll. She commented on

---

* The ancient world had clung just as fiercely to the code of secrecy. Legend has it that Pythagoras banished one of his followers (or in some accounts threw him off a boat, drowning him) for "telling men who were not worthy" a dreadful mathematical secret. Hippasus's sin was revealing to outsiders the discovery that certain numbers (in this case, the square root of 2) cannot be written down precisely ($^{14}/_{10}$ is close, for instance, but *no* fraction is exact). The Greeks found this numerical truth horrifying, a rip in the cosmic fabric.

how beautiful the stars were. Houtermans puffed out his chest. "I've known since yesterday why it is that they shine."

And no one else did. *That* was the point. Before the Royal Society proposed changing the rules, scientists had tried to have it both ways—they announced their discoveries, which let the world know they had solved a stubborn equation or designed a new clock mechanism or found the ideal shape for an arch, but often they concealed the details in a cipher, to be decoded only if someone else challenged the claim. The new call for full disclosure meant an about-face.

Hooke fought the call for openness with all his might, and he was not alone. Such resistance was practical as well as philosophical. Unlike Boyle, a man of enormous wealth, Hooke had a living to earn. He needed not simply to demonstrate his inventions but to patent them so he could turn a profit. For decades Hooke argued that the Royal Society ought to recast itself as a tiny army, like the conquistadors who had taken over Mexico. (He reserved the role of Cortez for himself.) Secrecy was vital, censorship of discoveries essential. "Nothing considerable in that kind can be obtained without secrecy," Hooke warned, "because else others not qualified . . . will share of the benefit."

Hooke lost that battle, but his doubts highlight just how radical the new approach was. In the past, scholars and intellectuals had always made a point of setting themselves apart from the common herd, and they had invoked biblical authority to justify themselves. "Do not throw your pearls before swine," they intoned endlessly, "lest they trample them under foot and turn to attack you." Like other priesthoods, intellectuals had long luxuriated in arcane rites and obscure vocabulary. The new scientists could have taken the same line. That would have seemed a natural step and an endorsement of a deeply entrenched and hugely powerful doctrine—true knowledge was too deep to put in ordinary words and too dangerous to trust to ordinary mortals.

Astonishingly, they did just the opposite. Rather than set themselves up as the newest mystic brotherhood, the new scientists spearheaded an attack on exclusivity. This marked just as sharp a break with the past as the attack on secrecy. In the era when science was born, carrying out experiments and building instruments still looked suspiciously like manual labor. That was not a way to win admirers. In the past, the discovery of truth had always been a task reserved for philosophers. Now technicians and tinkerers wanted to horn in.*

The renown that the Royal Society eventually won makes it easy to forget just how shaky its triumph was. The very sweep of its innovations made its survival doubtful. In its early decades, the Society never managed to establish itself as a safe, permanent feature of the intellectual landscape. More than once it nearly went under, beset by financial woes or bad leadership or personality clashes. For that reason, for long stretches it will nearly vanish from our story.

On this question of practicality, Hooke could scarcely have made his distaste for the old ways more clear. Universities might still believe that educating their students meant equipping them to compose odes in Greek and epigrams in Latin. Hooke favored a different mission. Even across the centuries his voice drips

---

* We still see relics of that prejudice against "applied" knowledge today. The historian Paolo Rossi notes that the term "liberal arts" originally came into use to mark off those areas of study deemed proper for a gentleman's education. These were the fields suited to free men (*liberi*) rather than to servants or slaves.

with scorn. The aim of science was "to improve the knowledge of natural things and all useful Arts . . . not meddling with Divinity, Metaphysics, Morals, Politics, Grammar, Rhetorick, or Logick."

The disdain was aimed not at learning but at endless talking. (Hooke was the furthest thing from a philistine. Architect, scientist, inventor—"England's Leonardo," in one biographer's phrase—he had set out as a young man to become an artist.)* But Hooke and his restless allies had work to do, and they were in a hurry to get started. They sought to carry out their investigations "not by a glorious pomp of words," one early manifesto declared, "but by the silent, effectual, and unanswerable arguments of real productions."

This was a battle cry, too, though again we might miss its significance. The rejection of "glorious" phrasemaking was a deliberate provocation. The seventeenth century was an age of tremendous formality, especially when it came to speech and writing. The Royal Society would have none of it. The Society favored "a close, naked, natural way of speaking," its first historian declared, ". . . bringing all things as near the Mathematical plainness as they can, and preferred the language of artisans, countrymen, and merchants before that of wits or scholars."

This was shocking. To speak in a "naked, natural way" was as unlikely as to walk outdoors naked and naturally. Elaborate rules of etiquette governed every kind of verbal exchange. A person sitting down to write a letter had to know when it was

---

* As a thirteen-year-old, Hooke briefly apprenticed with the famous portrait painter Peter Lely. (It was Lely whom Oliver Cromwell instructed to "paint my picture truly like me," warts and all.) Hooke's artistic career came to an early end when he found he was allergic to the paints and oils in Lely's studio.

proper to sign "Your most obedient and most obliged servant" and when "Your most humble and most affectionate servant." If the letter was addressed to a social superior, eloquent groveling was mandatory. "All that I mean," John Donne wrote to the Duke of Buckingham, "in using this boldness, of putting myself into your Lordship's presence by this rag of paper, is to tell your Lordship that I lie in a corner, as a clod of clay, attending what kind of vessel it shall please you to make of Your Lordship's humblest and thankfullest and devotedst servant."

In books even such arcane matters as the precise appearance of the dedication page called for great concern. Such pages carried a fervent declaration of praise and gratitude from the author to his patron. The size of the blank space between dedication and the author's signature was key. The larger the gap in status between patron and author, the larger the gap between dedication and signature, as if to ensure that the unkempt, ink-stained writer could not besmirch his eminent sponsor.

Such rules endured all through the 1600s, but the Royal Society set out to combat them. Metaphors, similes, and all the other long-esteemed forms of verbal display were mere distractions, ornamental froufrou that only impeded the search for truth. Out with them!

Chapter Twelve
DOGS AND RASCALS

The changes took decades to play out, but the contours of the new landscape took shape early on. Thomas Hobbes, the philosopher, had seen the new world coming even before the founding of the Royal Society. Informal though it was, the Society grew out of a series of even more haphazard gatherings of various experimentalists. In 1655, Hobbes had cast his lot with the new scientists. He invited all men to pursue truth as scientists did, by spelling out their reasoning in ordinary language and by carrying out experiments in public. The method was open to everyone. "If you would like," Hobbes assured his readers, "you too can use it."

This was a democratic idea in a world deeply mistrustful of democracies. But something had shifted, and Hobbes had spotted it. Dry-as-dust scholarship in musty archives was out, independent investigation in. Pedigree was beside the point; so were Latin quotations; so were the opinions of ancient authors. Science was a game that anyone could play, which meant that everything was up for grabs. Anyone could propose a new idea, and no idea was exempt from challenge. This is the sense in which the scientific revolution was indeed revolutionary.

Nonetheless, even many who fought on the revolutionary side harbored doubts about the program. Isaac Newton, for one, recoiled at the thought of catering to ordinary, educated readers. He never revealed his writings on alchemy, and though he did publish his greatest work, on gravity, he took enormous trouble to move it as far as humanly possible from anyone's notion of a "natural way of speaking." Newton published his masterpiece, *Philosophiae Naturalis Principia Mathematica* (*The Mathematical Principles of Natural Philosophy*), in the form of an enormously long mathematical argument. Theorem, proof, and corollary follow one another in stately procession as in the world's most difficult geometry textbook, the austere work unleavened by a word of guidance or explanation. The tone throughout is one of "glacial remoteness," one modern physicist observes, and "makes no concessions to the reader."

Many great mathematicians are nearly as hard to follow as Newton. Disdainful of those stumbling after them, they take as their motto Samuel Johnson's remark that "I have found you an argument, I am not obliged to find you an understanding."* Sometimes the motive for presenting work in its finished, polished state is aesthetic, akin to an artist's careful rubbing out of the grid lines that helped him get his proportions right. But not in Newton's case. He had "designedly made his *Principia* abstruse," he wrote, so that he would not be "baited by little Smatterers in Mathematics." What others could not grasp, they could not criticize. Those capable of following his reasoning would see its merits.

But Newton belonged with the rebels despite his hostility to them. By temperament the least open of men, it was his ironic fate to advance science so dramatically that new recruits, inspired by his example, came flooding in. The new generation of scientists spoke in ordinary language and published their findings for

---

* The esteemed eighteenth-century mathematician Laplace, for example, inspired despair even in his admirers. "I never came across one of Laplace's 'Thus it plainly appears,'" wrote one, "without feeling sure that I have hours of hard work before me to fill up the chasm and find out and show how it plainly appears."

all to read. They thought they were paying homage to Newton, who would have hated them.

The new approach brought a torrent of progress, but progress had a price. Science became a race run in public, and the first across the line hoisted the trophies. The Royal Society started the first-ever scientific journal, *Philosophical Transactions* (now in its fourth century). In 1672 the *Transactions* published a hugely important article, Newton's report that "pure" white light contains within itself all the colors of the spectrum. The paper, almost as much as the discovery itself, marked a breakthrough. This was, the historian I. Bernard Cohen observed, "the first time that a major scientific discovery was announced in print in a periodical."

From now on, journals and books would trumpet the news of discoveries and hail the innovators' genius. The victors won fame and honor. Everyone else was left to sulk and snipe. Many of the early scientists, as it happened, were bad-tempered, ferociously competitive men, which only raised the stakes. And in these early days, no rules of combat had yet arisen. In time, for instance, scientists would establish a system of peer review as the gold standard in their field. Before a reputable journal published a paper, a team of expert, independent, anonymous referees would have to deem it new and significant.

Even today, with such structures long established, science is a contact sport. Early on, the scrambling was far fiercer. Scientific jobs were rare, and self-promotion was an essential skill. Even great scientists had to fit their scientific work into the nooks and crannies of their day, around their "real" jobs as clergymen or doctors or diplomats, or they had to woo princes or other deep-pocketed patrons. Artists and writers had long known the dubious pleasures of patronage. Now scientists learned the same lessons. Patrons tended to be fickle and quickly bored, charmed by wit but put off by rigor.

Making matters worse, science seemed a field designed to stir up feuds. Writers and artists no doubt felt as much hostility toward one another as scientists did, but they had an easier time going different ways. Ben Jonson didn't have to write a play about a Scottish king and his scheming wife. Science was a race to a single goal. Ready, set, go! Build a clock that works even on a ship careening in ten-foot waves. Find a way to explain why Saturn looks so strange through a telescope. Take a few scattered observations and compute the shape of a comet's path.

For each question, one winner, many losers. Rivals shouted insults at one another or fumed in silence. Feuds burned on for decades. Isaac Newton and John Flamsteed, the first royal astronomer, hated one another. Newton warred with Hooke, too, and Hooke despised Newton in return, as well as Christiaan Huygens, the great Dutch astronomer, and a dozen more. Hooke denounced his enemies as "dogs," "raskalls," and "spies" who had stolen ideas that rightfully belonged to him. Newton and Gottfried Leibniz abused one another with terms that made Hooke's insults sound loving.

"If I have seen farther than others," Newton once remarked, "it is because I have stood on the shoulders of giants." That famous declaration, usually cited as one of Newton's rare ventures into generosity, was not quite the tribute it appears. Newton's aim was evidently to praise various of his forebears but also to mock his enemy Hooke, a slight, twisted figure far closer to a hunchback than a giant.

"Nullius in Verba" may have been the Royal Society's official motto, but the Society's members were only intermittently high-minded. They would all have understood Gore Vidal's remark that "it is not enough to succeed. Others must fail."

Chapter Thirteen

A DOSE OF POISON

This was a callous era, both in everyday life and in science. Weakness inspired scorn, not pity. Blindness, deafness, a club-foot, or a twisted leg were rebukes from God. Entertainments were often cruel, punishments invariably brutal, scientific experiments sometimes macabre. For decades, for example, dissections had been performed in public for ticket-buying audiences, like plays in a theater. The bodies of executed criminals made ideal subjects for study and display and not simply because they were readily available. Just as important, one historian notes, cutting criminals open in front of an attentive audience demonstrated "the culture's preference for punishment by means of public humiliation and display."

That preference was on display year-round. When it comes to punishing wrongdoers, modern society tends to avert its eyes. Not so the 1600s. In London prisoners locked in the pillory provided a bit of street theater, an alternative to a puppet show. Passersby screamed insults or took the opportunity to show their children what happened to bad people. The captive stood upright as best he could, head and hands trapped in holes cut into a horizontal wooden beam. Perhaps his ears had been nailed to the beam. The pillory was built to pivot as the prisoner staggered, in order to give spectators on all sides a chance to throw a dead cat or a rock.

Since punishments were meant to frighten and demean, whippings, brandings, and hangings took place where crowds could gather. Thieves could be hanged for stealing a handkerchief, though that was rare. More often, the theft of a handkerchief or a parcel of bread and cheese brought a whipping. A bolder theft—a gold ring or a silver bracelet—might merit branding with a hot iron, with a *T* for *thief*. Usually the *T* was seared into the flesh of the hand, although for a brief era that was considered too lenient, and the cheek was used instead. Any substantial theft meant death on the gallows.

Religious dissenters risked terrible punishments, like criminals. For the sin of "horrid blasphemy," in 1656, the Quaker James Nayler was sentenced to three hundred lashes, the branding of a *B* on his forehead, and the piercing of his tongue with a red-hot iron. Then Nayler was flung into prison, where he served three years in solitary confinement.

Even the most gruesome tortures served as spectacle and entertainment. (One history of seventeenth-century London includes an outing to watch a hanging in a section titled "Excursions.") The most dreadful punishment of all was hanging, drawing, and quartering. "A man sentenced to this terrible fate was strung up by the neck, but not so as to kill him," the historian Liza Picard explains. "Then his innards were taken out as if he were a carcass in a butcher's shop. This certainly killed him, if he had not died of shock before. The innards were burned, and the eviscerated corpse was chopped into four bits, which with the head were nailed up here and there throughout the City." (To preserve severed heads so that they could endure years of outdoor exposure, and to keep ravens away, they were parboiled with salt and cumin seeds.)

London Bridge in 1616, with traitors' heads on spikes above gateway (right foreground). The heads were such an everyday feature of life that the artist did not bother to call attention to them. By permission of the Folger Shakespeare Library.

London Bridge, more or less the shopping mall of its day, had been adorned for centuries with traitors' heads impaled on spikes. In Queen Elizabeth's day the bridge's southern gate bristled with some thirty heads.*

A taste for the grisly ran through the whole society, from the lowliest tradesman to the king himself. On May 11, 1663, Pepys made a passing reference to the king in his diary. Surgeons "did dissect two bodies, a man and a woman, before the King," Pepys wrote matter-of-factly, "with which the King was highly pleased."

At times the king's interest in anatomy grew downright creepy. At a court ball in 1663, a woman miscarried. Someone

---

* "As one of [Thomas More's] daughters was passing under the bridge," according to John Aubrey, "looking on her father's head, said she, 'That head has lain many a time in my lap, would to God it would fall into my lap as I pass under.' She had her wish, and it did fall into her lap, and is now preserved in a vault in the cathedral church at Canterbury."

brought the fetus to the king, who dissected it. To modern ears, the lighthearted tone surrounding the whole episode is almost unfathomable. "Whatever others think," the king joked, "he [i.e., Charles himself] hath the greatest loss . . . that hath lost a subject by the business."

When it came to experiments on animals, the seventeenth century was even less squeamish. Newton veered toward vegetarianism—he seldom ate rabbit and some other common dishes on the grounds that "animals should be put to as little pain as possible"—but such qualms were rare. Sages of the Royal Society happily carried out experiments on dogs that are too grim to read about without flinching. They had ample company. Descartes, as deep and introspective a thinker as ever lived, wrote blithely that humans are the only animals who think and feel. The yelp of a kicked dog no more indicated pain than did the sound of a drum when you beat it.

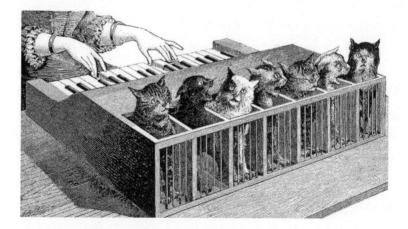

Another widely admired philosopher of the day, Athanasius Kircher, described an odd invention called a cat piano. The goal was to amuse a despondent prince. A row of cats sat in side-by-side cages, arranged according to the pitch of their meows. When the pianist pressed a key, a spike stabbed into the tail of the appropriate cat. "The result was a melody of meows that became more vigorous as the cats became more desperate. Who could help but laugh at such music? Thus was the prince raised from his melancholy."

In London shouting, jostling crowds flocked to bear-baitings and bull-baitings, where they could watch a chained animal fight a pack of slavering dogs. (Thus the origin of the English bulldog, whose flat face and sunken nose let it keep its hold on a flailing bull without having to open its powerful jaws to breathe.) Even children's games routinely featured the torment of animals. "No wonder," the historian Keith Thomas writes, "that traditional nursery rhymes portray blind mice having their tails cut off with a carving knife, blackbirds in a pie, and pussy in the well."

Experiments on dogs were considered entertaining as well as informative. Wren, for instance, made a specialty of splenecto-mies, surgical operations to remove the spleen. With a dog tied in place on a table, Wren would carefully cut into its abdomen, extract the spleen, tie off the blood vessels, sew up the wound, and then place the poor beast in a corner to recover, or not. (Boyle subjected his pet setter to the procedure and noted that the dog survived "as sportive and wanton as before.")

The operations provide yet another instance of how new sci-ence and ancient belief found themselves yoked together. For fourteen centuries, the Western world had endorsed Galen's doctrine that health depended on a balance of four "humors"— blood, phlegm, yellow bile, and black bile—each secreted by a different organ.* Too little or too much phlegm, say, made a

---

* The word *disease* is a relic of this theory. When the humors fell out of balance, the patient's *ease* gave way to *dis-ease.*

person *phlegmatic*, dreary and sluggish and flat. Just as the heart was the source of blood, so the spleen was the source of black bile (which, in the wrong proportion, caused melancholy). All medical authorities had so decreed for more than a thousand years. Hence Wren's experiment, a new test of an age-old dogma—if health depended on having all four humors in the proper balance, what would it mean if a dog could get along perfectly well with no bile-producing spleen at all?

Countless dogs suffered through transfusions, too. Many of them survived, somehow, even though no one knew about the dangers of infection or mismatched blood types. Boyle wrote a paper calling for answers to such questions "As whether a *fierce* Dog, by being often quite new stocked with the blood of a *cowardly* Dog, may not become more tame," or "whether a Dog, taught to fetch and carry, or to dive after Ducks, or to sett, will after frequent and full recruits of the blood of Dogs unfit for those Exercises, be as good at them, as before?"

Sometimes the experiments had more serious rationales. How, for instance, did venom from a snakebite spread throughout the body? What about a person who swallowed poison? What would happen if someone injected him with poison instead? Tempting as it might have been to test such ideas on human "volunteers," dogs came first. (Boyle did report a conversation with "a foreign Ambassador, a very curious person," who had set out to inject one of his servants with poison. The servant spoiled the fun by fainting before the experiment could begin.)

But many of the experiments were essentially stunts. At dinner one November night in 1666, Pepys listened to an excited report of the events a few days before at the Royal Society. Dr. William Croone gave a vivid account of a blood transfusion between a mastiff and a spaniel. "The first died upon the place," Pepys reported, "and the other very well, and likely to do well."

Croone had been impressed by the "pretty experiment" and even suggested to Pepys that someday transfusions might prove useful "for the amending of bad blood by borrowing from a better body." But no one at the Royal Society had dwelt much on the medical significance of the day's entertainment. The mood had been carefree, the company devoting most of its attention to a kind of parlor game. Which natural enemies would make the most amusing partners for a blood exchange? "This did give occasion to many pretty wishes," Pepys wrote cheerily, "as of the blood of a Quaker to be let into an Archbishop, and such like."

Chapter Fourteen

OF MITES AND MEN

Pepys's light tone was telltale. Science was destined to remake the world, but in its early days it inspired laughter more often than reverence. Pepys was genuinely fascinated with science—he set up a borrowed telescope on his roof and peered at the moon and Jupiter, he raced out to buy a microscope as soon as they came on the market, he struggled through Boyle's *Hydrostatical Paradoxes* ("a most excellent book as ever I read, and I will take much pains to understand him through if I can"), and in the 1680s he served as president of the Royal Society—but his amusement was genuine, too.* All these intellectuals studying spiders and tinkering with pumps. It *was* a bit ludicrous.

The king certainly thought so. He, too, was an aficionado of science. He had, after all, chartered the Royal Society, and he liked to putter about in his own laboratory. But he referred

---

* Like James Thurber, who never managed to see anything through a microscope but a reflection of his own eye, Pepys had trouble getting the hang of his microscope. "My wife and I with great pleasure," he wrote in his diary in August 1664, "but with great difficulty before we could come to find the manner of seeing anything."

to the Society's savants as his "jesters," and once he burst out laughing at the Royal Society "for spending time only in weighing of ayre, and doing nothing else since they sat."

Weighing the air—which plainly weighed nothing at all—seemed less like a groundbreaking advance than a return to such medieval pastimes as debating whether Adam had a navel. Skeptics never tired of satirizing scientists for their impracticality. One critic conceded that the members of the Royal Society were "Ingenious men and have found out A great Many Secrets in Nature." Still, he noted, the public had gained "Little Advantage" from such discoveries. Perhaps the learned scientists could turn their attention to "the Nature of butter and cheese."

In fact, they had given considerable thought to cheese, and also to finding better ways to make candles, pump water, tan leather, and dye cloth. From the start, Boyle had taken the lead in speaking out against any attempts to separate science and technology. "I shall not dare to think myself a true naturalist 'til my skill can make my garden yield better herbs and flowers, or my orchard better fruit, or my field better corn, or my dairy better cheese" than the old ways produced.

To hear the scientists and their allies tell it, unimaginable bounty lay just around the corner. Joseph Glanvill, a member of the Royal Society but not a scientist himself, shouted the loudest. "Should those Heroes go on, as they have happily begun," Glanvill exclaimed, "they'll fill the world with *wonders*." In the future, "a voyage to Southern unknown Tracts, yea possibly the Moon, will not be more strange than one to America. To them that come after us, it may be as ordinary to buy a *pair of wings* to fly into remotest Regions, as now a pair of Boots to ride a Journey." *

Such forecasts served mainly to inspire the mockers. By 1676 the Royal Society found itself the subject of a hit London comedy, the seventeenth-century counterpart of a running gag on *Saturday Night Live*. The play was called *The Virtuoso*, which could mean either "far-ranging scholar" or "dilettante." Thomas Shadwell, the playwright, lifted much of his dialogue straight from the scientists' own accounts of their work.

Playgoers first encountered the evening's hero, Sir Nicholas Gimcrack, sprawled on his belly on a table in his laboratory. Sir Nicholas has one end of a string clenched in his teeth; the other end is tied to a frog in a bowl of water. The virtuoso's plan is to learn to swim by copying the frog's motions. A visitor asks whether he has tested the technique in water. Not necessary, says Sir Nicholas, who explains that he hates getting wet. "I content myself with the speculative part of swimming. I care not for the practical. I seldom bring anything to use. . . . Knowledge is my ultimate end."

Sir Nicholas's family is not pleased. A niece complains that he has "spent £2000 in Microscopes, to find out the nature of Eels in vinegar, Mites in Cheese, and the blue of Plums." A second

---

* Glanvill provides yet another example of how seventeenth-century scientists simultaneously endorsed new beliefs and clung to old ones. He argued strenuously in favor of science's new findings and at the same time insisted that spirits, demons, and witches were real. To deny the existence of evil spirits, Glanvill insisted, was to veer dangerously near to saying that only the tangible was real, and *that* was tantamount to atheism. No witches, no God!

niece worries that her uncle has "broken his Brains about the nature of Maggots and studied these twenty Years to find out the several sorts of Spiders."

All the favorite Royal Society pastimes came in for ridicule. Gimcrack studied the moon through a telescope, as Hooke had done, and his description of its "Mountainous Parts and Valleys and Seas and Lakes," as well as "Elephants and Camels," spoofs Hooke's account. (Hooke went to see the play and complained that the audience, which took for granted that he was the inspiration for Gimcrack, "almost pointed" at him in derision.)

Sir Nicholas experimented on dogs, too, and boasted about a blood transfusion in which "the *Spaniel* became a *Bull-Dog*, and the *Bull-Dog* a *Spaniel*." He had even tried a blood transfusion between a sheep and a madman. The sheep died, but the madman survived and thrived, except that "he bleated perpetually, and chew'd the Cud, and had Wool growing on him in great Quantities."

Like his king, Shadwell found much to satirize in the virtuosos' fascination with the properties of air. Sir Nicholas keeps a kind of wine cellar with bottles holding air collected from all over. His assistants have crossed the globe "bottling up Air, and weighing it in all Places, sealing the Bottles Hermetically." Air from Tenerife is the lightest, that from the Isle of Dogs heaviest. Shadwell had great fun with the notion that air is a substance, with properties, rather than a mere absence. "Let me tell you, Gentlemen," Sir Nicholas assures his visitors, "Air is but a thinner sort of Liquor, and drinks much the better for being bottled."

Shadwell had a good number of allies among the satirists of his day, many of them eminent. Samuel Butler lampooned men who spent their time staring into microscopes at fleas and drops of pond water and contemplating such mysteries as "How many different Species / Of Maggots breed in rotten Cheeses."

But no one brought as much talent to ridiculing science as Jonathan Swift. Even writing more than half a century after the founding of the Royal Society, in *Gulliver's Travels*, Swift quivered with indignation at scientists for their pretension and impracticality. (Swift visited the Royal Society in 1710, squeezing in his visit between a trip to the insane asylum at Bedlam and a visit to a puppet show.)

Gulliver observes one ludicrous project after another. He sees men working on "softening Marble for Pillows and Pincushions" and an inventor engaged in "an Operation to reduce human Excrement to its original Food." In many places, the satire targets actual Royal Society experiments. Real scientists had struggled in vain, for instance, to sort out the mysterious process that would later be called photosynthesis. How do plants manage to grow by "eating" sunlight?* Gulliver meets a man who "had been Eight Years upon a project for extracting Sun-Beams out of Cucumbers, which were to be put into Vials hermetically sealed, and let out to warm the Air in raw inclement Summers."

Swift's sages live in the expectation that soon "one Man shall do the Work of Ten and a Palace may be built in a Week," but none of the high hopes ever pans out. "In the mean time, the whole Country lies miserably waste, the Houses in Ruins, and the People without Food or Cloaths."

Mathematicians, the very emblem of head-in-the-clouds uselessness, come in for extra ridicule. So absentminded are they

* The mystery would only be unraveled around 1800.

that they need to be rapped on the mouth by their servants to remember to speak. Lost in thought, they fall down the stairs and walk into doors. They can think of nothing but mathematics and music. Even meals feature such mathematical courses as "a Shoulder of Mutton, cut into an Equilateral Triangle; a Piece of Beef into a Rhomboides; and a Pudding into a Cycloid."

In hardheaded England, where "practicality" and "common sense" were celebrated as among the highest virtues, Swift's disdain for mathematics was widely shared by his fellow intellectuals. In that sense, Swift's mockery of absentminded professors was standard issue. But, more than he could have known, Swift was right to direct his sharpest thrusts at mathematicians. These dreamers truly were, as Swift intuited, the most dangerous scientists of all. Microscopes and telescopes were the glamorous innovations that drew all eyes—*Gulliver's Travels* testifies to Swift's fascination with their power to reveal new worlds—but new instruments were only part of the story of the age. The insights that would soon transform the world required no tools more sophisticated than a fountain pen.

For it was the mathematicians who invented the engine that powered the scientific revolution. Centuries later, the story would find an echo. In 1931, with great hoopla, Albert Einstein and his wife, Elsa, were toured around the observatory at California's Mount Wilson, home to the world's biggest telescope. Someone told Elsa that astronomers had used this magnificent telescope to determine the shape of the universe. "Well," she said, "my husband does that on the back of an old envelope."

Those outsiders who did take science seriously tended to dislike what they saw. The scientists themselves viewed their work as a way of paying homage to God, but their critics were not so sure. Astronomy stirred the most fear. Who needed it, when we already know the story of the heavens and the Earth, and on the best possible authority? To probe further was to treat the Bible as just another source of information, to be tested and questioned like any other. A popular bit of seventeenth-century doggerel purportedly captured the scientists' view: "All the books of Moses / Were nothing but supposes."

The devout had another objection. Science diverted its practitioners from deep questions to silly ones. "Is there anything more Absurd and Impertinent," one minister snapped, "than to find a Man, who has so great a Concern upon his Hands as the preparing for Eternity, all busy and taken up with *Quadrants*, and *Telescopes*, *Furnaces*, *Syphons*, and *Air Pumps?*"

So science irritated those who found it pompous and ridiculous. It offended those who found it subversive. Just as important, it bewildered almost everyone.

Chapter Fifteen

A PLAY WITHOUT
AN AUDIENCE

The new science inspired ridicule and hostility partly for the simple reason that it *was* new. But the resentment had a deeper source—the new thinkers proposed replacing a time-honored, understandable, commonsense picture of the world with one that contradicted the plainest facts of everyday life. What could be less disputable than that we live on a fixed and solid Earth? But here came a new theory that *began* by flinging the Earth out into space and sending it hurtling, undetectably, through the cosmos. If the world is careening through space like a rock shot from a catapult, why don't we feel it? Why don't we fall off?

The goal of the new scientists—to find ironclad, mathematical laws that described the physical world in all its changing aspects—had not been part of the traditional scientific mission. The Greeks and their successors had confined their quest for perfect order to the heavens. On Earth, nothing so harmonious could be expected. When the Greeks looked to the sky, they saw the sun, the moon, and the planets moving imperturbably on their eternal rounds.* The planets traced complicated paths (*planet* is Greek for "wanderer"), but they continued on their way, endlessly. On the corrupt Earth, on the other hand, all motions were short-lived. Drop a ball and it bounces, then rolls, then stops. Throw a rock and seconds later it falls to the ground. Then it sits there.

Ordinary objects could certainly be *set* moving—an archer tensed his muscles, drew his bow, and shot an arrow; a horse strained against its harness and pulled a plow—but here on Earth an inanimate body on its own would not *keep* moving. The archer or the horse evidently imparted a force of some kind, but whatever that force was it soon dissipated, as heat dissipates from a poker pulled from a fire.

Greek physics, then, began by dividing its subject matter into two distinct pieces. In the cosmos above, motion represents the natural state of things and goes on forever. On the Earth below, *rest* is natural and motion calls for an explanation. No one saw this as a problem, any more than anyone saw a problem in different nations following different laws. Heaven and Earth completely differ from one another. The stars are gleaming dots of light moving across the sky, the Earth a colossal rock solid and immobile at the center of the universe. The heavens are predictable, the Earth anything but. On June 1, to pick a date at random, we know what the stars in the night sky will look like, and we know that they will look virtually the same again

---

* The moon gave the Greeks problems. It was a heavenly body, which meant it had to be perfect and unblemished, but no one could miss its patches of light and dark. One attempted explanation: the moon was a perfect mirror and its dark spots were the reflections of oceans on Earth.

on June 1 next year, and next century, and next millennium.* What June 1 will bring on Earth this year, or any year, no one knows.

Aristotle had explained how it all works, both in the heavens and on Earth, about three hundred years before the birth of Christ. For nearly two thousand years everyone found his scheme satisfactory. All earthly objects were formed from earth, air, fire, and water. The heavens were composed of a fifth element or essence, the *quintessence*, a pure, eternal substance, and it was only in that perfect, heavenly domain that mathematical law prevailed. Why do everyday, earthly objects move? Because everything has a home where it belongs and where it returns at the first opportunity. Rocks and other heavy objects belong down on the ground, flames up in the air, and so on. A "violent" motion—flinging a javelin into the air—might temporarily overcome a "natural" one—the javelin's impulse to fall to the ground—but matters quickly sort themselves out.

The picture made sense of countless everyday observations: Hold a candle upright or turn it downward, and the flame rises regardless. Hoist a rock overhead in one hand and a pebble in the other, and the rock is harder to hold aloft. Why? Because it is bigger and therefore more earth-y, more eager to return to its natural home.

---

* The stars will not look exactly the same, mostly because the earth wobbles a bit on its axis, like a spinning top. But the changes are so small that art historians and astronomers, working together, have answered such questions as what the sky over St.-Rémy-de-Provence looked like on June 19, 1889, the night Van Gogh painted "Starry Night." (Van Gogh stuck remarkably close to reality.)

All such explanations smacked of biology, and to modern ears the classical world sounds strangely permeated with will and desire. Why do falling objects accelerate? "The falling body moved more jubilantly every moment because it found itself nearer home," writes one historian of science, as if a rock were a horse returning to the barn at the end of the day.

The new scientists would strip away all talk of "purpose." In the new way of thinking, rocks don't *want* to go anywhere; they just fall. The universe has no goals. But even today, though we have had centuries to adapt to the new ideas, the old views still exert a hold. We cannot help attributing goals and purposes to lifeless nature, and we endlessly anthropomorphize. "Nature abhors a vacuum," we say, and "water seeks its own level." On a cold morning we talk about the car starting "reluctantly" and then "dying," and if it just won't start we pound the dashboard in frustration and mutter, "Don't do this to me."

It was Galileo more than any other single figure who finally did away with Aristotle. Galileo's great coup was to show that for once the Greeks had been too cautious. Not only were the heavens built according to a mathematical plan, but so was the ordinary, earthly realm. The path of an arrow shot from a bow could be predicted as accurately as the timing of an eclipse of the sun.

This was a twofold revolution. First, the kingdom of mathematics suddenly claimed a vast new territory for itself. Second, all those parts of the world that could *not* be described mathematically were pushed aside as not quite worthy of study. Galileo made sure that no one missed the news. Nature is "a book written in mathematical characters," he insisted, and anything that could not be framed in the language of equations was "nothing but a name."*

Aristotle had discussed motion, too, but not in a mathematical way. *Motion* referred not only to change in position, which can easily be reduced to number, but to every sort of change—a ship sailing, a piece of iron rusting, a man growing old, a fallen tree decaying. Motion, Aristotle decreed in his *Physics*, was "the actuality of a potentiality." Galileo sneered. Far from investigating the heart of nature, Aristotle had simply been playing word games, and obscure ones at that.

In the new view, which Galileo hurried to proclaim, the scientist's task was to describe the world objectively, as it really is, not subjectively, as it appears to be. What was objective—tangible, countable, measurable—was real and primary. What was subjective—the tastes and textures of the world—was dubious and secondary. "If the ears, the tongue, and the nostrils were taken away," wrote Galileo, "the figures, the numbers, and the motions would indeed remain, but not the odors nor the tastes nor the sounds."

This was an enormous change. Peel away the world of appearances, said Galileo, and you find the real world beneath. The world consists exclusively of particles in motion, pool balls colliding on a vast table. All the complexity around us rises out of that simplicity.

After Galileo and Newton, the historian of science Charles C.

---

* Galileo's intellectual offspring espouse the same view today, in virtually identical words. "To those who do not know mathematics it is difficult to get across a real feeling as to the beauty, the deepest beauty, of nature," wrote the physicist Richard Feynman. "If you want to learn about nature, to appreciate nature, it is necessary to understand the language that she speaks in."

Gillispie has written, science would "communicate in the language of mathematics, the measure of quantity," a language "in which no terms exist for good or bad, kind or cruel . . . or will and purpose and hope." The word *force*, for example, Gillispie noted, "would no longer mean 'personal power' but 'mass-times-acceleration.'"

That austere, geometric world has a beauty of its own, Galileo and all his intellectual descendants maintained. The problem is that most people cannot grasp it. Mathematicians believe fervently that their work is as elegant, subtle, and rich as any work of music. But everyone can appreciate music, even if they lack the slightest knowledge of how to read a musical score. For outsiders to mathematics—which is to say, for almost everyone—advanced mathematics is a symphony played out in silence, and all they can do is look befuddled at a stage full of musicians sawing away to no apparent effect.

The headphones that would let everyone hear that music do exist, but they can only be built one pair at a time, by the person who intends to wear them, and the process takes years. Few people take the trouble. In the centuries that followed the scientific revolution, as the new worldview grew ever more dominant, poets would howl in outrage that scientists had stripped the landscape bare. "Do not all charms fly / At the mere touch of cold philosophy?" Keats demanded. Walt Whitman, and many others, would zero in even tighter. "When I heard the learn'd astronomer," wrote Whitman, the talk of figures, charts, and diagrams made him "tired and sick."

Mankind had long taken its place at the center of the cosmos for granted. The world was a play performed for our benefit. No longer. In the new picture, man is not the pinnacle of creation but an afterthought. The universe would carry on almost exactly the same without us. The planets trace out patterns in the sky, and those patterns would be identical whether or not humans had ever taken notice of them. Mankind's role in the cosmic drama is that of a fly buzzing around a stately grandfather clock.

The shift in thinking was seismic, and the way it came about had nothing in common with the textbook picture of progress in science. Change came not from finding new answers to old questions but from abandoning the old questions, unanswered, in favor of new, more fruitful ones. Aristotle had asked *why*. Why do rocks fall? Why do flames rise? Galileo asked *how*. How do rocks fall—faster and faster forever, or just until they reach cruising speed? How fast are they traveling when they hit the ground?

Aristotle's *why* explained the world, Galileo's *how* described it. The new scientists began, that is, by dismissing the very question that all their predecessors had taken as fundamental. (Modern-day physicists often strike the same impatient tone. When someone asked Richard Feynman to help him make sense of the world as quantum mechanics imagines it, he supposedly snapped, "Shut up and calculate.")

Aristotle had an excellent answer to the question *why do rocks fall when you drop them?* Galileo proposed not a different answer or a better one, but no answer at all. People do not "know a thing until they have grasped the 'why' of it," Aristotle insisted, but Galileo would have none of it. To ask why things happen, he declared, was "not a necessary part of the investigation."

And that change was only the beginning.

Chapter Sixteen

ALL IN PIECES

Galileo, Newton, and their fellow revolutionaries immediately turned their backs on yet another cherished idea. This time they banished common sense. Long acquaintance with the world had always been hailed as the surest safeguard against delusion. The new scientists rejected it as a trap. "It is not only the heavens that are not as they seem to be, and not only motion," Descartes argued, in a modern historian's paraphrase. "The whole universe is not as it seems to be. We see about us a world of qualities and of life. They are all mere appearances."

It was a Polish cleric and astronomer named Nicolaus Copernicus who had struck the first and hardest blow against common sense. Despite the evidence, plain to every child, that we live on solid ground and that the sun travels around us, Copernicus argued that everyone has it all wrong. The Earth travels around the sun, and it spins like a top as it travels. And no one feels a thing.

This was ludicrous, as everyone who heard about the new-fangled theory delighted in pointing out. For one thing, the notion of a sun-centered universe contradicted scripture. Had not Joshua ordered the sun (rather than the Earth) to stand still in the sky? This was a huge hurdle. In the 1630s, nearly a century after Copernicus's death, Galileo would face the threat of torture and then die under house arrest for arguing in favor of a sun-centered universe.

(Isaac Newton was born in the year that Galileo died. That was coincidence, but in hindsight it seemed to presage England's rise to scientific preeminence and Italy's long drift to mediocrity. What was not coincidence was that seventeenth-century England welcomed science, on the grounds that science supported religion, and thrived; and seventeenth-century Italy feared science, on the grounds that science undermined religion, and decayed.)

Copernicus himself had hesitated for decades before publishing his only scientific work, *On the Revolutions of the Celestial Spheres*, perhaps because he knew it would stir religious fury as well as scientific opposition. Legend has it that he was handed the first copy of his masterpiece on his deathbed, on May 24, 1543, although by that point he may have been too weak to recognize it.

Religion aside, the scientific objections were enormous. If Copernicus was right, the Earth was speeding along a gigantic racetrack at tens of thousands of miles an hour, and none of the passengers suffered so much as a mussed hair. The fastest that *any* traveler had ever moved was roughly twenty miles an hour, on horseback.

These arguments came from the most esteemed scholars, not from yokels. They knew, on both scientific and philosophical grounds, that the Earth does not move. (Aristotle had argued that the Earth rests in place because it occupies its natural home, the center of the universe, just as an ordinary object on the ground stays in *its* place unless something comes along and dislodges it.) Scholars pointed to countless observations that all led to the same conclusion. We can be sure the Earth stands still, one eminent philosopher explained, "for at the slightest jar of the Earth, we would see cities and fortresses, towns and mountains thrown down."

But we don't see cities toppled, the skeptics noted, nor do we see any other evidence that we live on a hurtling platform. If we're racing along, why can we pour a drink into a glass without worrying that the glass will have moved hundreds of yards out of range by the time the drink reaches it? If we climb to the roof and drop a coin, why does it land directly below where we let it go and not miles away?

But Copernicus's new doctrine inspired fear as well as ridicule and confusion, because it led almost at once to questions that transcended science. If the Earth was only one planet among many, were those other worlds inhabited, too? By what sort of creatures? Had Christ died for *their* sins? Did they have their own Adam and Eve, and what did that say about evil and original sin? "Worst of all," in the words of the historian of science Thomas Kuhn, "if the universe is infinite, as many of the later Copernicans thought, where can God's Throne be located? In an infinite universe, how is man to find God or God man?"

Copernicus could not disarm such fears by pointing to new discoveries or new observations. He never looked through a telescope—Galileo would be one of the first to turn telescopes to the heavens, some seven decades after Copernicus's death—and in any case telescopes could not *show* the Earth moving but only provided evidence that let one deduce its motion.

On the contrary, everything that Copernicus could see and feel spoke in favor of the old theories and against his own. "Sense pleads for Ptolemy," said Henry More, a colleague of Newton at Cambridge and a distinguished English philosopher. But common sense lost out. The old, Earth-centered theory that Ptolemy had devised was a mathematical jumble, and that marked it for death. The old system worked perfectly well, but it was a hodgepodge.

The great challenge to pre-Copernican astronomy had to do with sorting out the motions of the planets, which do not trace a simple course through the sky but at some point interrupt their journey and loop back in the direction they've just come from. (The stars present no such mystery. Each night Greek astronomers watched them rotating smoothly through the sky, turning in a circle with the North Star at its center. Each constellation moved around the center, like a group of horses on a merry-go-round, but the stars within a constellation never rearranged themselves.)

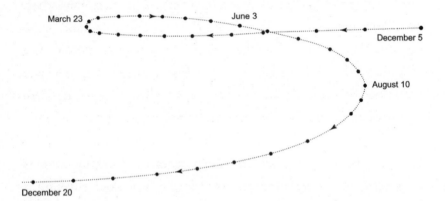

The path of Saturn as seen from Earth, as depicted by Ptolemy in 132–33 A.D. From March through June, Saturn appears to reverse course.

Accounting for the planets' strange course changes would have been enough to give classical astronomers fits. Making the challenge all the harder, classical doctrine decreed that planets must travel in circular orbits (since planets are heavenly objects and circles are the only perfect shape). But circular orbits didn't fit the data. The solution was a complicated mathematical dodge in which the planets traveled not in circles but in the next best thing—in circles attached to circles, like revolving seats on a Ferris wheel, or even in circles attached to circles attached to circles.

Copernicus tossed out the whole complicated system. The planets weren't really moving sometimes in one direction and sometimes in the other, he argued, but simply orbiting the sun. The reason those orbits look so complicated is that we're watching from the Earth, a moving platform that is itself circling the sun. When we pass other planets (or they pass us), it looks as if they've changed course. If we could look down on the solar system from a vantage point above the sun, all the mystery would vanish.

This new system was conceptually tidier than the old one, but it didn't yield new or better predictions. For any practical question—predicting the timing of eclipses and other happenings in the solar system—the old system was fully as accurate as the new. No wonder Copernicus kept his ideas to himself for so long. And yet think of the astonishing leap this wary thinker finally nerved himself to make. With no other rationale but replacing a cumbersome theory with one that was mathematically more elegant, he dared *to set the Earth in motion*.

A few intellectuals might have been won over by a revolutionary argument with nothing in its favor but aesthetics. Most people wanted more. How did the new theory deal with the most basic questions? "If the moon, the planets and comets were of the same nature as bodies on earth," wrote Arthur Koestler, "then they too must have 'weight'; but what exactly does 'the weight' of a planet mean, what does it press against or where does it tend to fall? And if the reason why a stone falls to Earth is not the Earth's position in the center of the universe, then just why does the stone fall?"

Copernicus did not have answers, nor did he have anything to say about what keeps the planets in their orbits or what holds the stars in place. The Greeks *had* provided such answers, and the answers had stood for millennia. (Each planet occupied a spot on an immense, transparent sphere. The spheres were nested, one inside the other, and centered on the Earth. The stars occupied the biggest, most distant sphere of all. As the spheres turned, they carried the planets and the stars with them.)

No one could yet answer the new questions about the stars and planets. No one knew why objects on Earth obey one set of laws and bodies in the heavens another. No one even knew where to look for answers. John Donne, poet and cleric, spoke for many of his perplexed, frustrated contemporaries. "The Sun is lost, and th' earth, and no man's wit / Can well direct him where to look for it," he lamented, in a poem written a year after Galileo first looked through his telescope.

"The new Philosophy calls all in doubt," Donne wrote in another verse. "'Tis all in pieces, all coherence gone."

Part Two

HOPE AND MONSTERS

Chapter Seventeen

NEVER SEEN UNTIL THIS MOMENT

Virginia Woolf famously remarked that "on or about December 1910 human character changed." She might have picked a different date, almost precisely three centuries earlier. On January 7, 1610, Galileo turned a telescope to the night sky. Human nature—or at least humankind's picture of the universe and our own place within it—changed forever.

Three months later Galileo told the world what he had seen, in a book called *The Starry Messenger*. On the day the book reached Venice, the English ambassador, Sir Henry Wotton, sent a startled letter home. "I send herewith unto his Majesty the strangest piece of news (as I may justly call it) that he hath ever yet received from any part of the world." Sir Henry's emphasis on the word *news* was fitting. What he was about to pass on was not merely "news" in the modern, journalistic sense but "news" in the truest sense—a report of something that until that moment had never been seen or even imagined.

What was this astonishing news? "The Mathematical Professor at Padua . . . hath discovered four new planets rolling about the sphere of Jupiter"—four new planets in the unchanging heavens—and that was only part of the story. Galileo had also uncovered the secret of the Milky Way; he had learned, for the first time, the true nature of the moon, with all its pockmarked imperfections; he had found that the supposedly pristine sun was marred with black spots. In short, Wotton reported in slack-jawed astonishment, "he hath . . . overthrown all former astronomy."

Four decades before, the Danish astronomer Tycho Brahe had startled the world with a discovery of his own. In 1572 Tycho saw what he took to be a new star in the constellation Cassiopeia.* The last of the great naked-eye astronomers, Tycho was a meticulous observer with an unsurpassed knowledge of the sky. He had known "all the stars in the sky" from boyhood, he boasted, but even casual stargazers knew Cassiopeia, with its striking W shape. The supposed star shone so brightly that it could be seen during the day. It stayed in view for well over a year, which meant that it couldn't be a comet. It never changed its position against the backdrop of other stars, which meant that it had to be immensely far away. Nothing but a star had those properties. It was undeniable, and it was impossible.

Today every promising actor or athlete is a "new star," and the cliché has lost its force, but the appearance of the *first* new star in the immutable heavens was shocking. Tycho proclaimed it "the greatest wonder that has ever shown itself in the whole of nature since the beginning of the world, or in any case as great as when the Sun was stopped by Joshua's prayers."

Unable to sort out its meaning, most observers labeled this

---

* Modern-day astronomers have shown that Tycho's star was a supernova, an exploding star, rather than a new one.

aberration "Tycho's star" and did their best to put it out of their minds. But in 1604, still another new star appeared, this one perhaps even brighter than its predecessor. Galileo, caught up in the excitement, delivered a public lecture on the new star to a standing-room-only crowd. The discovery of two new stars within three decades shocked the learned world. Stargazers knew the appearance of the night sky as intimately as coast dwellers know the sea. We miss the point if we downplay their astonishment. How could a star appear where no star could be? All Europe was as stunned as another group, on the other side of the Atlantic, at almost the same moment.

On the morning of September 3, 1609, a band of Indians fishing from dugout canoes just off present-day Manhattan saw something odd in the distance. At first, it was only clear that the strange object was "something remarkably large swimming or floating on the water, and such as they had never seen before." These first witnesses raced to shore and recruited reinforcements. The object drew closer. The guesswork grew frenzied, "some [of the Indians] concluding it either to be an uncommon large fish or other animal, while others were of opinion it must be some very large house." The mysterious object drew closer still and then halted, its huge white wings billowing. In fear and fascination, the Indians on shore and the sailors on the deck of Henry Hudson's *Half Moon* stood staring at one another.

What was it like to see what no one had ever seen before?

In that same year of 1609, perhaps in May, Galileo heard talk of a Dutch invention, a lens maker's device with the power to bring far-off objects into close view. By this time, reading glasses to compensate for farsightedness were centuries old. Glasses to help with nearsightedness were more recent but widely available, too. Lenses for farsightedness were convex, thick in the middle and thin at the edges (lentil-shaped, hence the word *lens*); lenses for nearsightedness were concave, thinner in the middle than at the edges. The breakthrough that made the telescope possible was to combine a convex lens with a concave one. Everything hinged on the proportion between the strengths of the two lenses, which called for difficult feats of grinding and polishing.

By the end of August, Galileo had built one of the sorcerer's tubes for himself. It didn't look like much—a skinny tube about a yard long made mostly from paper and wood, it resembled a tightly rolled poster—and it took a bit of fiddling to get the hang of seeing through it. Galileo unveiled it to a group of high-ranking Venetians. They took turns peering through his telescope and responded with "infinite amazement," in Galileo's proud words.

"Many of the nobles and senators, although of a great age, mounted more than once to the top of the highest church tower in Venice," Galileo reported, "in order to see sails and shipping that were so far off that it was two hours before they were seen, without my spy-glass, steering full sail into the harbor." The military advantages of such an invention were plain, but Galileo made sure that no one could miss them. The telescope, he pointed out, allows its users "to discover at a much greater distance than usual the hulls and sails of the enemy, so that for two hours and more we can detect him before he detects us."

Galileo rocketed to fame. Thrilled by what they had seen, the senators immediately doubled his salary and awarded him a lifetime contract at Padua. (Galileo had helped his own cause by presenting the senators an elaborate telescope as a gift, this one no drab tube but an ornate instrument in red and brown leather decorated, like an elegant book, in gold filigree.)

Galileo's decision to highlight the telescope's value for warfare and commerce was cagey, but it was necessary, too. Galileo had grand ambitions. He knew from the start that the real discoveries would come from looking up to the stars, not out to sea. Which meant that the world had to be cajoled into believing that it could trust the sights revealed by this new, mysterious invention. In Rome, in 1611, he pointed his telescope at a palace far in the distance, and "we readily counted its each and every window, even the smallest." With the telescope trained on a distant wooden sign, "we distinguished even the periods carved between the letters."

So the telescope provided honest information. It *revealed* true features of faraway objects; it didn't somehow, through trickery or strange properties of light and lenses, conjure up mirages. If Galileo had simply aimed his telescope at the heavens, without preliminaries, skeptics might have dismissed the wonders he claimed to see. (Even so, some people refused to look, as today some might shy away from a purported ESP machine.)

Galileo continued to improve his design and soon produced a telescope able to magnify twenty times, twice as powerful as his first model. He could now be certain that the new stars that had appeared in 1572 and 1604 were the merest prelude, a two-note introduction to a visual symphony.

His own excitement fully matched that of the ecstatic senators in San Marco's belltower. The "absolute novelty" of his discoveries, Galileo wrote, filled him with "incredible delight." He marveled at the sight of "stars in myriads, which have never been seen before, and which surpass the old, previously known stars in number more than ten times." The moon, the perfect disc of a thousand poets' odes, "does not possess a smooth and polished surface but one rough and uneven and just like the face of the Earth itself, everywhere full of vast protuberances, deep chasms, and sinuosities."

The Milky Way was not some kind of cosmic fog reflecting light from the sun or moon, as had long been speculated. A glance through the telescope would at once put an end to all "wordy disputes upon this subject," Galileo boasted, and would leave no doubt that "the Galaxy is nothing else but a mass of innumerable stars planted together in clusters . . . many of them tolerably large and extremely bright, and the number of small ones quite beyond determination."

Yet another discovery was the most important of all, guaranteed to "excite the greatest astonishment by far." But even for Galileo the astonishment was slow in dawning. He had aimed his telescope at Jupiter and spotted several bright objects near the planet. The next day they were still there, but they had rearranged themselves. A few days later, another rearrangement. Some days there were four objects; some days only two or three. What could it mean?

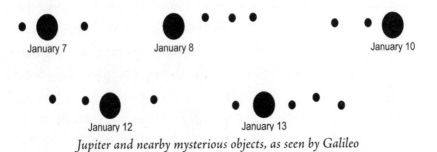

Jupiter and nearby mysterious objects, as seen by Galileo

The answer, Galileo came to see, was that four objects were in orbit around Jupiter. "I have discovered four planets, neither known nor observed by any one of the astronomers before my time," Galileo crowed. (He hurried to name the moons in honor of Cosimo de' Medici, Grand Duke of Tuscany, who swallowed the bait with gusto.) Here was a planetary system in miniature, and not a diagram or a mathematical hypothesis but an observable reality. Jupiter's moons were mini-Earths moving in orbit around a central body. Why could not the Earth itself be in orbit around a huge central body? And if the Earth, why not the other planets, too?

These were exhilarating, disorienting discoveries. The unsuspected vistas revealed by the telescope inspired a fair number of seventeenth-century thinkers to rejoice at this proof that God's creation was truly without bounds. It was only fitting that an infinite God should have created an infinite universe. What could be "more splendid, glorious, and magnificent than for God to have made the universe commensurate with his own immensity?" asked the Royal Society's Joseph Glanvill.

The gates to the cosmos had been thrown open, and optimists ran through and turned cartwheels in the vastness. "When the heavens were a little blue arch stuck with stars, methought the universe was too strait and close," exulted the French writer Bernard de Fontenelle, in an immensely popular account of the new doctrines called *On the Plurality of Worlds.* "I was almost stifled for want of air; but now it is enlarged in height and breadth and a thousand vortexes taken in. I begin to breathe with more freedom, and I think the universe to be incomparably more magnificent than it was before."*

---

* Fontenelle's exuberance was characteristic, though consistency was not his strength. In the next breath he professed, with equal verve, to be worried that the immensity of the universe made his own efforts seem tiny and irrelevant. Like Carl Sagan in our day, he was known as much for enthusiasm as scholarship. Fontenelle lived to be one hundred and scarcely slowed down along the way. Near the end, he met one famous beauty and remarked, "Ah madame, if I were only eighty again!"

But the endless expanses that beckoned so invitingly to some induced a kind of trembling agoraphobia in others. Pascal spoke for all who found themselves horrified by a vision of the planets as specks of dust adrift in a black immensity. "The eternal silence of these infinite spaces frightens me," he observed, and he seemed to view humankind on its lonely voyage as akin to a ship's crew adrift in an endless sea. "What is a man in the midst of infinity?" Pascal asked.

Decades before, Copernicus's pushing of the Earth off center stage had inspired similar questions and similar fears, but among a smaller audience. Galileo had far more impact. Anyone could look through a telescope, while almost no one could follow a mathematical argument. But whether Copernicus or Galileo took the role of narrator, the story was the same. The Earth was not the center of the universe but a run-of-the-mill planet in a random corner of the cosmos.

This stripping away of Earth's special status is always cited as a great assault on human pride. Freud famously contended, for example, that in the course of modern history three thinkers had dealt enormous blows to humankind's self-esteem. The three were Copernicus, Darwin, and Freud himself. Darwin had proved that humans are animals, and Freud that we are blind to our own motivations. But the first body blow had come from Copernicus, who had displaced mankind from his place of honor.

Freud had a key piece of the story almost exactly backward. Before Copernicus and Galileo, humans *had* believed that they lived at the center of the universe, but in their minds the center was a shameful, degraded place, not an exalted one. The Earth was lowly in every sense, at the furthest possible remove from the heavens. Man occupied "the filth and mire of the world," wrote

Montaigne, "the worst, lowest, most lifeless part of the universe, the bottom story of the house."

In the cosmic geography of the day, heaven and hell were actual places. Hell was not consigned to some vague location "below" but sat deep inside the Earth with its center directly beneath Jerusalem. The Earth was the center of the universe, and hell was the center of the center. Galileo's adversary Cardinal Bellarmine spelled out why that was so. "The place of devils and wicked damned men should be as far as possible from the place where angels and blessed men will be forever. The abode of the blessed (as our adversaries agree) is Heaven, and no place is further removed from Heaven than the center of the earth."

Mankind had always occupied a conspicuous place in the universe, in other words, but it was a dangerous and exposed position rather than a seat of honor. Theologians through the ages had thought well of that arrangement precisely because it did *not* puff up human pride. Humility was a virtue, they taught, and a home set amid "filth and mire" was nearly certain to have humble occupants.

In a sense, Copernicus had done mankind a favor. By moving the Earth to a less central locale, he had moved humankind farther from harm's way. For religious thinkers, ironically, this was yet another reason to object to the new doctrine. Theologians found themselves contemplating a riddle—how to keep humanity in its place when its place had moved?

In time, they would come up with an answer. They would seize on a different aspect of the new astronomy, the vast expansion in the size of the universe. If the universe was bigger, then man was smaller. For theologians in search of a way to reconcile themselves to science's new teachings, a doctrine that seemed to belittle mankind was welcome news.

Chapter Eighteen

FLIES AS BIG AS A LAMB

The microscope came along a bit later than the telescope, but its discovery produced just as much amazement. Here, too, were new worlds, and this time teeming with life! The greatest explorer of these new kingdoms was an unlikely conquistador, a Dutch merchant named Antonie van Leeuwenhoek. He seems to have begun tinkering with lenses with no grander ambition than to check for defects in swatches of cloth.

Leeuwenhoek quickly moved beyond fabric samples. Peering through microscopes he built himself—more like magnifying glasses than what we think of as microscopes—he witnessed scenes that no one else had ever imagined. In a frenzy of excitement, he dashed off letters to the Royal Society, hundreds altogether, describing the "secret world" he had found. He thrilled at the living creatures in a drop of water scooped from a puddle and then found he did not even have to venture outdoors to find teeming, complex life. He put his own saliva under the microscope and "saw, with great wonder, that in the said matter there were many very little living animalcules, very prettily a-moving. The biggest sort had a very strong and swift motion, and shot through the spittle like a pike does through the water."

Hooke had been experimenting with microscopes of his own

design for years. Leeuwenhoek's microscopes yielded clearer images, but on November 15, 1677, Hooke reported that he, too, had seen a great number of "exceedingly small animals" swimming in a drop of water. And he had witnesses. Hooke rattled off a list: "Mr. Henshaw, Sir Christopher Wren, Sir John Hoskyns, Sir Jonas Moore, Dr. Mapletoft, Mr. Hill, Dr. Croone, Dr. Grew, Mr. Aubrey, and diverse others." The roll call of names highlights just how shocking these findings were. The microscope was so unfamiliar, and the prospect of a tiny, living, hitherto invisible world so astonishing, that even an eminent investigator like Hooke needed allies. It would be as if, in our day, Stephen Hawking turned a new sort of telescope to the heavens and saw UFOs flying in formation. Before he told the world, Hawking might coax other eminent figures to look for themselves.

But Hooke and the rest of the Royal Society could not catch Leeuwenhoek. Endlessly patient, omnivorously curious, and absurdly sharp-eyed, he racked up discovery after discovery.* Sooner or later, everything—pond water, blood, plaque from his teeth—found its way to his microscope slides. Leeuwenhoek jumped up from his bed one night, "immediately after ejaculation before six beats of the pulse had intervened," and raced to his microscope. There he became the first person ever to see sperm cells. "More than a thousand were moving about in an amount of material the size of a grain of sand," he wrote in amazement, and "they were furnished with a thin tail, about five or six times

---

* Leeuwenhoek was a contemporary of Vermeer. Both men lived in Delft, the two shared a fascination with light and lenses, and Leeuwenhoek served as executor of Vermeer's will. Some art historians believe that Vermeer's *Astronomer* and his *Geographer* both depict Leeuwenhoek, but no one has been able to prove that Leeuwenhoek and Vermeer ever met.

as long as the body . . . and moved forward owing to the motion of their tails like that of a snake or an eel swimming in water."* Leeuwenhoek hastened to assure the Royal Society that he had obtained his sample "after conjugal coitus" (rather than "by sinfully defiling myself"), but he did not discuss whether Mrs. Leeuwenhoek shared his fascination with scientific observation.

No matter. Others did. Even Charles II delighted in peering through microscopes and witnessing life in miniature. "His Majesty seeing the little animals, contemplated them in astonishment and mentioned my name with great respect," Leeuwenhoek wrote proudly. This was a development almost as striking as Leeuwenhoek's findings themselves. In the new world of science, a merchant who had never attended a university and knew only Dutch, not Latin, could make discoveries that commanded the attention of a king.

Both the microscope and the telescope fascinated the seventeenth century's intelligentsia, not just its scientists. The telescope tended to produce unwelcome musings on man's puniness, as we have seen, but the picture of worlds within worlds revealed by the microscope did not trouble most people. Pascal was an exception. The endless descent into microworlds—"limbs with joints, veins in these limbs, blood in these veins, humors in this blood, globules in these humors, gases in these globules"—left him queasy and afraid. Many a ten-year-old has delighted in an imaginary outward zoom that plays Pascal's voyage in reverse: *I live at 10 Glendale Road in the town of Marblehead in the county of Essex in the state of Massachusetts in the United States of America*

* The microscope that Leeuwenhoek used on that fateful night was put up for auction in April 2009. The winning bidder paid $480,000.

on the planet Earth in the Milky Way galaxy. Pascal's inward journey shared the same rhythm, but the dread in his tone stood the child's exhilaration on its head.

Most people felt more fascination than fright, perhaps simply because we tend to feel powerful in proportion to our size. In any case, both telescope and microscope strengthened the case for God as designer. The ordinary world had already provided countless examples of God's craftsmanship. "Were men and beast made by fortuitous jumblings of atoms," Newton wrote contemptuously, "there would be many parts useless in them— here a lump of flesh, there a member too much." Now the microscope showed that God had done meticulous work even in secret realms that man had never known. Unlike those furniture makers, say, who lavished all their care on the front of their bureaus and desks but neglected surfaces destined to stay hidden, God had made *every* detail perfect.

The heavens declared the glory of God, and so did fleas and flies and feathers. Man-made objects looked shoddy in comparison. Hooke examined the tip of a needle under a microscope, to test the aptness of the expression "as sharp as a Needle." He found not a perfect, polished surface but "large Hollows and Roughnesses, like those eaten in an Iron Bar by Rust and Length of Time." A printed dot on the page of a book told the same story. To the naked eye it looked "perfectly black and round," wrote Hooke, "but through the Magnifier it seemed grey, and quite irregular, like a great Splatch of London Dirt."

No features of the natural world were too humble to inspire rapt study. In some of the earliest experiments with microscopes, Galileo had tinkered with various designs. His astonishment reaches us across a gap four centuries wide. Galileo had seen "flies which look as big as a lamb," he told a French visitor, "and are covered all over with hair, and have very pointed nails by means of which they keep themselves up and walk on glass, although hanging feet upwards."

Many of the objects that came in for close examination were even less grand than houseflies. In April 1669 Hooke and the other members of the Royal Society gazed intently at a bit of fat and then at a moldy smear of bookbinder's paste, "which was found to have a fine moss growing on it." One early scientist who studied plants under the microscope marveled that "one who walks about with the meanest stick holds a piece of nature's handicraft which far surpasses the most elaborate . . . needlework in the world."

Hooke published a lavish book called *Micrographia* that featured such stunning illustrations (by Hooke himself) as a twelve-by-eighteen-inch foldout engraving of a flea. The creature

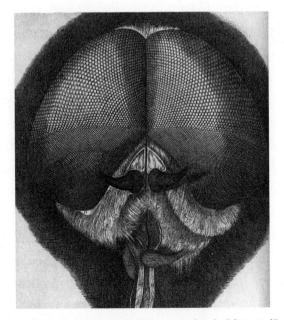

Robert Hooke's drawing of a fly's meticulously "designed" eyes

was, Hooke noted admiringly, "adorn'd with a curiously polish'd suit of sable Armour, neatly jointed." Another oversize illustration showed a fly's eyes, with some fourteen thousand facets or "pearls." Hooke went out of his way to justify lavishing attention on so lowly an insect. "There may be as much curiosity of contrivance and structure in every one of these Pearls, as in the eye of a Whale or Elephant," he wrote, and he noted that in any case God was surely up to such a task. "As one day and a thousand years are the same with him, so may one eye and ten thousand."

Both telescope and microscope had opened up new worlds. The new vistas served to reinforce the belief that on every scale the universe was a flawless, harmonious, and unimaginably complex mechanism. God was a sculptor who could shape stars and planets and a craftsman with a delicacy of touch to shame the finest jeweler.

Chapter Nineteen

FROM EARTHWORMS
TO ANGELS

If the thinkers of the seventeenth century had been content to see God as a superbly talented artist and craftsman, their homage might have taken a different form. Instead they looked at the marvelous sights revealed by the telescope and microscope and found new support for their favorite doctrine, that God was a mathematician.

They believed it already, thanks largely to their discoveries about the geometry of the cosmos, but they saw the new evidence as proving the case beyond the least possible doubt. In part this was because of the new sights themselves. Seen through the microscope, the least imposing objects revealed a geometer's shaping hand. One early scientist wrote an astonished hymn to grains of salt, which turned out to be "Cubes, Rhombs, Pyramids, Pentagons, Hexagons, Octagons" rendered "with a greater Mathematical Exactness than the most skilful Hand could draw them."

But the renewed emphasis on God-the-mathematician came mostly by way of a different, stranger path. One of the seventeenth century's most deeply held beliefs had to do with the so-called great chain of being. The central idea was that all the objects that had ever been created—grains of sand, chunks of gold, earthworms, lions, human beings, devils, angels—occupied a particular rank in a great chain that extended from the lowest of the low to the hem of God's garment. Nearby ranks blended almost insensibly into one another. Some fish had wings and flew into the air; some birds swam in the sea.

It was a fantastically elaborate system, though it strikes modern ears as more akin to a magical realist fantasy than a guide to everyday life. Purely by reasoning, the intellectuals of the seventeenth century believed, they could draw irrefutable conclusions about the makeup of the world. Angels, for example, were as real as oak trees. Since God himself had fashioned the great chain, it was necessarily perfect and could not be missing any links. So, just as there were countless creatures reaching *downward* from humans to the beasts, there had to be countless steps leading *upward* from humans to God. QED.

That made for a lot of angels. "We must believe that the angels are there in marvelous and inconceivable numbers," one scholar wrote, "because the honor of a king consists in the great crowd of his vassals, while his disgrace or shame consists in their paucity. Thousands of thousands wait on the divine majesty and tenfold hundreds of millions join in his worship."

Each link had its proper place in the hierarchy, king above noble above commoner, husband above wife above child, dog above cat, worm above oyster. The lion was king of beasts, but *every* domain had a "king": the eagle among birds, the rose among flowers, the monarch among humans, the sun among the stars. The various kingdoms themselves had specific ranks, too, some lower and some higher—stones, which are lifeless, ranked lower than plants, which ranked lower than shellfish, which ranked lower than mammals, which ranked lower than angels, with innumerable other kingdoms filling all the ranks in between.

In a hierarchical world, the doctrine had enormous intuitive appeal. Those well placed in the pecking order embraced it, unsurprisingly, but even those stuck far from the top made a virtue of "knowing one's place." Almost without exception, scholars and intellectuals endorsed the doctrine of the all-embracing, immutable great chain. To say that things might be different was to suggest that they could be better. This struck nearly everyone as both misguided—to attack the natural order was to shake one's fist at the tide—and blasphemous. Since God was an infinitely powerful creator, the world necessarily contained all possible things arranged in the best possible way. Otherwise He might have done more or done better, and who would presume to venture such a criticism?

As usual, Alexander Pope summarized conventional wisdom in a few succinct words. No one ever had less reason to endorse the status quo than Pope, a hunchbacked, dwarfish figure who lived in constant pain. He strapped himself each day into a kind of metal cage to hold himself upright. Then he took up his pen and composed perfectly balanced couplets on the theme that God has His reasons, which we limited beings cannot fathom. "Whatever is, is right."

The great chain had a long pedigree, and from the beginning the idea that the world was jam-packed had been as important as the idea that it was orderly. Plato had decreed that "nothing incomplete is beautiful," as if the world were a stamp album and any gap in the collection an affront. By the 1600s this view had long since hardened into dogma. If it was possible to do something, God would do it. Otherwise He would be selling himself short. Today the cliché has it that we use only 10 percent of our brains. For a thousand years philosophers and naturalists wrote as if to absolve God from that charge. "The work of the creator would have been incomplete if aught could be added to it," one French scientist declared blithely. "He has made all the vegetable species which could exist. All the minute gradations of animality are filled with as many beings as they can contain."

This was also the reason, thinkers of the day felt certain, that God had created countless stars and planets where the naked eye saw only the blackness of space. God had created infinitely many worlds, one theologian and Royal Society member explained, because only a populous universe was "worthy of an infinite CREATOR, whose *Power* and *Wisdom* are without bounds and measures."

But why did that all-powerful creator have to be a mathematician? Gottfried Leibniz, the German philosopher who took all knowledge as his domain, made the case most vigorously. The notion of a brim-full universe provided Leibniz the opening he needed. Leibniz was as restless as he was brilliant, and, perhaps predictably, he believed in an exuberantly creative God. "We must say that God makes the greatest number of things that he can," Leibniz declared, because "wisdom requires variety."

Leibniz immediately proceeded to demonstrate his own wisdom by making the same point in half a dozen varied ways. Even if you were wealthy beyond measure, Leibniz asked, would you choose "to have a thousand well-bound copies of Virgil in your library"? "To have only golden cups"? "To have all your buttons made of diamonds"? "To eat only partridges and to drink only the wine of Hungary or of Shiraz"?

Now Leibniz had nearly finished. Since God loved variety, the only question was how He could best ensure it. "To find room for as many things as it is possible to place together," wrote Leibniz, God would employ the fewest and simplest laws of nature. *That* was why the laws of nature could be written so compactly and why they took mathematical form. "If God had made use of other laws, it would be as if one should construct a building of round stones, which leave more space unoccupied than that which they fill."

So the universe was perfectly ordered, impeccably rational, and governed by a tiny number of simple laws. It was not enough to *assert* that God was a mathematician. The seventeenth century's great thinkers felt they had done more. They had proved it.

The scientists of the 1600s felt that they had come to their view of God by way of argument and observation. But they were hardly a skeptical jury, and their argument, which seemed so compelling to its original audience, sounds like special pleading today. Galileo, Newton, Leibniz, and their peers leaped to the conclusion that God was a mathematician largely because *they* were mathematicians—the aspects of the world that intrigued them were those that could be captured in mathematics. Galileo found that falling objects obey mathematical laws and proclaimed that *everything* does. The book of nature is written in the language of mathematics, he wrote, "and the characters are triangles, circles and other geometrical figures, without whose help it is impossible to comprehend a single word of it; without which one wanders in vain through a dark labyrinth."

The early scientists took their own deepest beliefs and ascribed them to nature. "Nature is pleased with simplicity," Newton declared, "and affects not the pomp of superfluous causes." Leibniz took up the same theme. "It is impossible that God, being the most perfect mind, would not love perfect harmony," he wrote, and he and many others happily spelled out different features of that harmony. "God always complies with the easiest and simplest rules," Galileo asserted.

"Nature does not make jumps," Leibniz maintained, just as Einstein would later insist that "God does not play dice with the universe." We attribute to God those traits we most value. "If triangles had a god," Montesquieu would write a few decades later, "he would have three sides."

Newton and the others would have scoffed at such a notion. They were describing God's creation, not their own. Centuries later, a classically minded revolutionary like Einstein would still hold to the same view. In an essay on laws of nature, the mathematician Jacob Bronowski wrote about Einstein's approach to science. "Einstein was a man who could ask immensely simple questions," Bronowski observed, "and what his life showed, and his work, is that when the answers are simple too, then you hear God thinking."

For a modern-day scientist like Bronowski, this was a rhetorical flourish. Galileo, Newton, and the other great men of the seventeenth century could have expressed the identical thought, and they would have meant it literally.

Chapter Twenty

THE PARADE
OF THE HORRIBLES

When Galileo and Newton looked at nature, they saw simplicity. That was, they declared, God's telltale signature. When their biologist colleagues looked at nature, they saw endless variety. That was, they announced, God's telltale signature.

Each side happily cited one example after another. The physicists pointed out that as the planets circle the sun, for instance, they all travel in the same direction and in the same plane. The biologists presented their own eloquent case, notably in a large and acclaimed book titled *The Wisdom of God Manifested in the Works of Creation.* The "vast Multitude of different Sorts of Creatures" testified to God's merits, the naturalist John Ray argued, just as it would show more skill in a manufacturer if he could fashion not simply one product but "Clocks and Watches, and Pumps, and Mills and [Grenades] and Rockets."

Strikingly, no one saw any contradiction in the views of the two camps. In part this reflected a division of labor. The physicists focused on the elegance of God's aesthetics, the biologists on the range of His inventiveness. Both sides were bound by the shared conviction, deeper than any possible division, that God had designed every feature of the universe. For the physicists, that view led directly to the idea that God was a mathematician, and progress. For biologists, it led down a blind alley and made the discovery of evolution impossible.

Two centuries passed between Newton's theory of gravity and Darwin's theory of evolution. How could that be? Newton's work bristled with mathematics and focused on remote, unfamiliar objects like planets and comets. Darwin's theory of evolution dealt in ordinary words with ordinary things like pigeons and barnacles. "How extremely stupid not to have thought of that!" Thomas Huxley famously grumbled after first reading Darwin's *Origin of Species*. No one ever scolded himself for not beating Newton to the *Principia*.

The "easier" theory proved harder to find because it required abandoning the idea of God the designer. Newton and his contemporaries never for a moment considered rejecting the notion of design. The premise at the heart of evolution is that living creatures have inborn, random differences; some of those random variations happen to provide an advantage in the struggle for life, and nature favors those variations. That focus on randomness was unthinkable in the seventeenth century. Even Voltaire, the greatest skeptic of his day, took for granted that where there was a design, there was a designer. No thinker of that age, no matter how brilliant, could imagine an alternative. "It is natural to admit the existence of a God as soon as one opens one's eyes," Voltaire wrote. "It is by virtue of an admirable art that all the planets dance round the sun. Animals, vegetables, minerals— everything is ordered with proportion, number, movement. Nobody can doubt that a painted landscape or drawn animals are works of skilled artists. Could copies possibly spring from an intelligence and the originals not?"

Newton, blinded by his faith in intelligent design, argued in the same vein. In a world where randomness was a possibility, he scoffed, we'd be beset with every variety of jury-rigged, misshapen creature. "Some kinds of beasts might have had but one eye, some more than two."

The problem was not simply that for Newton and the others "randomness" conveyed all the horror of "anarchy." Two related beliefs helped rule out any possibility of a seventeenth-century Darwin. The first was the assumption that every feature of the world had been put there for man's benefit. Every plant, every animal, every rock existed to serve us. The world contained wood, the Cambridge philosopher Henry More explained, because otherwise human houses would have been only "a bigger sort of beehives or birds' nests, made of contemptible sticks and straw and dirty mortar." It contained metal so that men could assault one another with swords and guns, rather than sticks, as they enjoyed the "glory and pomp" of war.

The second assumption that blinded Newton and his contemporaries to evolution was the idea that the universe was almost brand-new. The Bible put creation at a mere six thousand years in the past. Even if someone had conceived of an evolving natural world, that tiny span of time would not have offered enough elbow room. Small changes could only transform one-celled creatures into daffodils and dinosaurs if nature had eons to work with. Instead, seventeenth-century scientists took for granted that trees and fish, men and women, dogs and flowers all appeared full-blown, in precisely the form they have today.

Two hundred years later, scientists still clung to the same idea. In the words of Louis Agassiz, Darwin's great Victorian rival, each species was "a thought of God."

Chapter Twenty-One

"SHUDDERING BEFORE THE BEAUTIFUL"

The seventeenth century's faith that "all things are numbers" originated in ancient Greece, like so much else. The Greek belief in mathematics as nature's secret language began with music, which was seen not as a mere diversion but as a subject for the most intense study. Music was the great exception to the general rule that the Greeks preferred to keep mathematics untainted by any connection with the everyday world.

Pluck a taut string and it sounds a note. Pluck a second string twice as long as the first, Pythagoras found, and the two notes are an octave apart. Strings whose lengths form other simple ratios, like 3 to 2, sound other harmonious intervals.* That insight, the physicist Werner Heisenberg would say thousands of

---

* As one of Pythagoras's followers told the tale, the story began when Pythagoras listened to the sound of hammering as he walked by a blacksmith's shop. As the blacksmith struck the same piece of iron with different hammers, some sounds were harmonious, others not. The key, Pythagoras found, was whether the weights of the hammers happened to be in simple proportion. A twelve-pound hammer and a six-pound hammer, for instance, produced notes an octave apart.

years later, was "one of the truly momentous discoveries in the history of mankind."

Pythagoras believed, too, that certain numbers had mystical properties. The world was composed of four elements because 4 was a special number. Such notions never lost their hold. Almost a thousand years after Pythagoras, St. Augustine explained that God had created the world in six days because 6 is a "perfect" number. (In other words, 6 can be written as the sum of the numbers that divide into it exactly: $6 = 1 + 2 + 3$.)*

The Greeks felt sure that nature shared their fondness for geometry. Aim a beam of light at a mirror, for example, and it bounces off the mirror at the same angle it made on its incoming path. (Every pool player knows that a ball hit off a cushion follows the same rule.)

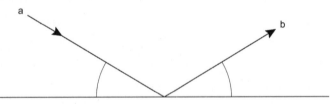

When light bounces off a mirror, the two marked angles are equal.

What looked like a small observation about certain angles turned out to have a big payoff—of the infinitely many paths that the light beam might take on its journey from *a* to a mirror to *b*, the path it actually does take is the shortest one possible. And there's more. Since light travels through the air at a constant speed, the *shortest* of all possible paths is also the *fastest*.

Even if light obeyed a mathematical rule, the rule might have

---

* Augustine did not explain why God did not make the world in 28 days ($1 + 2 + 4 + 7 + 14$) or 496 days or various other possibilities.

been messy and complicated. But it wasn't. Light operated in the most efficient, least wasteful way possible. This was so even in less straightforward circumstances. Light travels at different speeds in different mediums, for instance, and faster in air than in water. When it passes from one medium to another, it bends.

Look at the drawing below and imagine a lifeguard at *a* rather than a flashlight. If a lifeguard standing on the beach at *a* sees a person drowning at *b*, where should she run into the water? It's tricky, because she's much slower in the water than on land. Should she run straight toward the drowning man? Straight to a point at the water's edge directly in front of the flailing man?

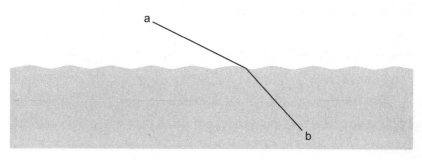

Light bends as it passes from air into water.

Curiously, this riddle isn't in the least tricky for light, which "knows" exactly the quickest path to take. "Light acts like the perfect lifeguard," physicists say, and over the centuries they've formulated a number of statements about nature's efficiency, not just to do with light but far more generally. The eighteenth-century mathematician who formulated one such principle proclaimed it, in the words of the historian Morris Kline, "the first scientific proof of the existence and wisdom of God."

Light's remarkable behavior was only one example of the seventeenth century's favorite discovery, that if a mathematical idea was beautiful it was virtually guaranteed to be useful. Scientists ever since Galileo and Newton have continued to find mysterious mathematical connections in the most unlikely venues. "You must have felt this, too," remarked the physicist Werner Heisenberg, in a conversation with Einstein: "the almost frightening simplicity and wholeness of the relationships which nature suddenly spreads out before us and for which none of us was in the least prepared."

For the mathematically minded, the notion of glimpsing God's plan has always exerted a hypnotic pull. The seduction is twofold. On the one hand, delving into the world's mathematical secrets gives a feeling of having one's hands on nature's beating heart; on the other, in a world of chaos and disaster, mathematics provides a refuge of eternal, unchallengeable truths and perfect order.

The intellectual challenge is immense, and the difficulty of the task makes the pursuit even more obsessive. In Vladimir Nabokov's novel *The Defense*, Aleksandr Luzhin is a chess grand master. He speaks of chess in just the way that mathematicians think of their field. While pondering a move and lighting a cigarette, Luzhin accidentally burns his fingers. "The pain immediately passed, but in the fiery gap he had seen something unbearably awesome—the full horror of the abysmal depths of chess. He glanced at the chessboard, and his brain wilted from unprecedented weariness. But the chessmen were pitiless; they held and absorbed him. There was horror in this, but in this also was the sole harmony, for what else exists in the world besides chess?"

Mathematicians and physicists share that passion, and unlike chess players they take for granted that they are grappling with nature's deepest secrets. (The theoretical physicist Subrahmanyan Chandrasekhar, a pioneer in the study of black holes, spoke of "shuddering before the beautiful.") They sustain themselves through the empty years with the unshakable belief that the answer is out there, waiting to be found. But mathematics is a cruel mistress, indifferent to the suffering of those who would woo her. Only those who themselves have wandered lost, wrote Einstein, know the misery and joy of "the years of searching in the dark for a truth that one feels but cannot express; the intense desire and the alternations of confidence and misgiving, until one breaks through to clarity and understanding."

The abstract truths that enticed Einstein and his fellow scientists occupy a realm separate from the ordinary world. That gulf between the everyday world and the mathematical one has, many times through the centuries, served as a lure rather than a barrier. When he was a melancholy sixteen-year-old, the modern-day philosopher and mathematician Bertrand Russell recalled many years later, he used to go for solitary walks "to watch the sunset and contemplate suicide. I did not, however, commit suicide, because I wished to know more of mathematics."

A deep dive into mathematics has special appeal, for it serves at the same time as a way to flee the world and to impose order on it. "Of all escapes from reality," the mathematician Gian-Carlo Rota observed, "mathematics is the most successful ever. . . . All other escapes—sex, drugs, hobbies, whatever—are ephemeral by comparison." Mathematicians have withdrawn from the dirty, dangerous world, they believe, and then, by thought alone, they have added new facts to the world's store of knowledge. Not just new facts, moreover, but facts that will stand forever, unchallengeable. "The certainty that [a mathematician's] creations will endure," wrote Rota, "renews his confidence as no other pursuit." It is heady, seductive business.

Perhaps this accounts for the eagerness of so many seventeenth-century intellectuals to look past the wars and epidemics all around them and instead to focus on the quest for perfect, abstract order. Johannes Kepler, the great astronomer, barely escaped the religious battles later dubbed the Thirty Years' War. One close colleague was drawn and quartered and then had his tongue cut out. For a decade his head, impaled on a pike, stood on public display next to the rotting skulls of other "traitors."

Kepler came from a village in Germany where dozens of women had been burned as witches during his lifetime. His mother was charged with witchcraft and, at age seventy-four, chained and imprisoned while awaiting trial. She had poisoned a neighbor's drink; she had asked a grave digger for her father's skull, to make a drinking goblet; she had bewitched a villager's cattle. Kepler spent six years defending her while finishing work on a book called *The Harmony of the World.* "When the storm rages and the shipwreck of the state threatens," he wrote, "we can do nothing more worthy than to sink the anchor of our peaceful studies into the ground of eternity."

Chapter Twenty-Two

PATTERNS MADE WITH IDEAS

For the Greeks, the word *mathematics* had vastly different associations than it does for most of us. Mathematics had almost nothing to do with adding up columns of numbers or figuring out how long it would take Bob and Tom working together to paint a room. The aim of mathematics was to find eternal truths—insights into the abstract architecture of the world—and then to prove their validity. "A mathematician, like a painter or poet, is a maker of patterns," wrote G. H. Hardy, an acclaimed twentieth-century mathematician and an ardent proponent of the Greek view. "If his patterns are more permanent than theirs, it is because they are made with ideas."

Let's take a few minutes to look at the kind of thing Greek mathematicians accomplished, because it was their example—and the way they interpreted their success—that inspired their intellectual descendants in the seventeenth century. (One of Newton's assistants could recall only one occasion when he had seen Newton laugh. Someone had made the mistake of asking Newton what use it was to study Euclid, "upon which Sir Isaac was very merry.") The Greeks had looked for their "permanent patterns" in the world of mathematics. Seventeenth-century scientists set out with the same goal except that they expanded their quest to the world at large.

They found mathematics on all sides. When Isaac Newton directed a beam of light through a prism, he marveled at the rainbow on his wall. No one could miss either the beauty or the order in that familiar spectacle, but it was the interplay between the two that so intrigued Newton. "A naturalist would scarce expect to see ye science of those colours become mathematicall," he wrote, "and yet I dare affirm that there is as much certainty in it as in any other part of Opticks."

For the Greeks the notion of "proof"—not a claim or a likelihood but actual proof beyond a doubt—was fundamental. A proof in mathematics is a demonstration or an argument. It starts with assumptions and moves, step by step, to a conclusion. But unlike ordinary arguments—who was the greatest president? who makes the best pizza in Brooklyn?—mathematical arguments yield irrefutable, permanent, universally acknowledged truths. *Of all the shapes you can make with a piece of string, a circle encloses the biggest area. The list of prime numbers never ends.* If three points aren't in a straight line, there is a circle that passes through all three.* Everyone who can follow the argument sees that it must be so.

Like other arguments, proofs come in many varieties. Mathematicians have individual, recognizable styles, just as compos-

---

* A prime number is one that can't be broken down into smaller pieces. For example, 2 is prime, and so are 3, 5, and 7; 10 is not prime (because 10 = 2 × 5). Prime numbers get rarer as you count higher and higher, but no matter how big a prime you name, there is always a bigger one.

ers and painters and tennis players do. Some think in pictures, others in numbers and symbols. The Greeks preferred to think pictorially. Take the Pythagorean theorem, for instance, perhaps the most famous theorem of them all. The theorem involves a right triangle—a triangle where one angle is 90 degrees—and relates the lengths of the various sides. In the simplest right triangle, one side is 3, another 4, and the longest 5. Many centuries before the birth of Christ, some unknown genius stared at those numbers—3, 4, 5—and saw something that astonished him.

It's easy to draw a triangle with a side 3 inches long and a side 4 inches long and a third side that's short (at left, below), or a triangle with a side 3 inches long and a side 4 inches long and a third side that's long (at right, below). But if the angle between the 3-inch side and the 4-inch one is not just any angle but 90 degrees, then the length of the third side turns out to be precisely 5. So the puzzle pieces that our unknown genius turned over and over in his mind were these: 3, 4, 5, 90 degrees. What tied those numbers together?

No doubt he drew endless right triangles and measured the sides. Nearly always the longest side would be a seemingly random number, no matter how carefully the two short sides were chosen. Even in the simplest case—a triangle where the two short sides were both 1 inch long—the third side didn't look simple at all. A shade more than 1⅜ inches, not even anything that lined up with the divisions on a ruler.

Perhaps he stuck with his experiments long enough to draw the right triangle with short sides 5 and 12. Set a ruler in place to draw the third side and then measure it. Success at last—the long side is precisely 13 inches long, so here is another right triangle with all three sides respectable whole numbers.

Two right triangles, two sets of numbers, like two words from a coded message. First set: 3, 4, 5. Second set: 5, 12, 13. What do the two triplets have in common?

For two thousand years, we have called the answer the Pythagorean theorem—the length of one short side, squared, plus the other short side, squared, equals the long side, squared. $3^2 + 4^2 = 5^2$. For the second triangle, $5^2 + 12^2 = 13^2$.* More to the point, the relationship holds for *every* right triangle whatsoever, whether it is scratched in the sand or splayed across the heavens.†

In modern terms, the theorem is usually written as $a^2 + b^2 = c^2$. In the pictorial terms the Greeks preferred, the theorem is about squares, not numbers, and Pythagoras's claim is that the area of one small square added to the area of the other small

---

* There are infinitely many choices of a, b, and c that satisfy $a^2 + b^2 = c^2$. But if you try *any* power higher than 2—if, for instance, you try to find whole numbers a, b, and c that satisfy $a^3 + b^3 = c^3$ or $a^4 + b^4 = c^4$—you will never find a single example that works (discounting the trivial case where a, b, and c are all set equal to 0). The statement that no such example exists is one of the most famous in mathematics. It is known as Fermat's last theorem, after the mathematician Pierre de Fermat, who jotted it down in the margin of a book in 1637. He had found "a truly marvelous proof," he scribbled, but "the margin is not large enough" to fit it. No one ever found his proof—presumably he'd made a mistake in his reasoning—and for more than three hundred years countless mathematicians tried and failed to find proofs of their own. Success finally came in 1995, as detailed in Amir Aczel's *Fermat's Last Theorem*.

† At the half-moon, for instance, sun, moon, and Earth form a right triangle.

square is exactly the same as the area of the large square. (See drawing below.) The two approaches, numerical and pictorial, are exactly equivalent. The choice between them is purely a matter of taste, like the choice between an architectural drawing and a scale model.

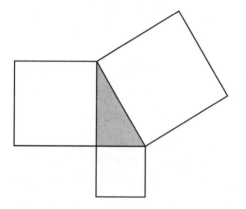

Pythagoras's theorem says that the area of one small square plus the area of the other small square is exactly equal to the area of the large square.

Chapter Twenty-Three

GOD'S STRANGE CRYPTOGRAPHY

If you had somehow happened to guess the Pythagorean theorem, how would you prove it? It's not enough to draw a right triangle, measure the sides, and do the arithmetic. That would only serve to verify the theorem for one example, not for *all* right triangles. Moreover, not even the most careful measuring could confirm that the sum worked out precisely, to the millionth decimal point and beyond, as it must. But even a dozen successful examples, or a hundred, or a thousand, would still fall short of proof. "True beyond a reasonable doubt" applies in law and in ordinary life—who doubts that the sun will rise tomorrow?—but the Greeks demanded more.

Here is one classic proof, which proceeds in jigsaw-puzzle fashion and almost wordlessly. In math as in chess, crucial moves often look mysterious at the time. Why move a knight *there* when all the action is *here*? In this case, the unexpected move that brings the answer in reach is this: take the original triangle and make three identical copies of it, so that you have four triangles all exactly the same.

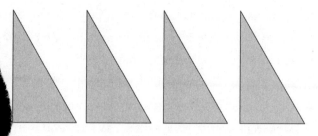

at's the gain in dealing with four triangles when we weren't
at to do with one? The gain comes in imagining the tri-
s cardboard cutouts and then sliding them around on a
different arrangements. Look at Figure X and Figure Y
Two different arrangements, both with the same four
s and some white space. In both cases, the outlines (in
ok like squares. How do we know they really are squares
t just four-sided, squarish shapes?

e at Figure X and Figure Y for a few seconds. All the bold
are the same length (because each bold side is made up
ong side of the original triangle and a short side). And all
corners are right angles. So the bold shape in Figure X is a
uare, and so is the bold shape in Figure Y, and both squares
re precisely the same size.

Almost done. Each bold square encloses the same area. Each
bold square is made up of four identical triangles and some white

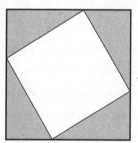
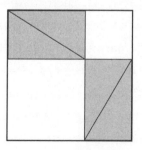

Figure X   *Figure Y*

space. Stare at the pictures again. The large white square in
Figure X *has to be* exactly the same in area as the two small
white squares in Figure Y. Voilà, we have Pythagoras!

Why did the Greeks find that discovery so astonishing?
its utility. No Greek would have asked, "What good is it?
good is a poem or a play? Would a sculpture be more ad
if it could also serve as a doorstop? Mathematics was tru
was beautiful, and that was more than enough. The po
not to find the length of a diagonal across a rectangul
without having to measure it, although the Pythagorea
rem lets you do that. The Greeks had loftier goals.

The Pythagorean theorem thrilled the Greeks for two
First, simply by thinking—without using any tools whatso
they had discovered one of nature's secrets, an eternal and
before-suspected truth about the structure of the world. S
they could prove it. Unlike nearly any other valid observati
vinegar is tart, Athens is dusty, Socrates is short—this partic
observation was not only true but *necessarily* true. One of Go
thoughts, finally grasped by man.

Like all the best insights, it is simultaneously inevitable and
surprising. But it may also be surprising that the Greeks took
for granted that their mathematical theorems were facts *about
the world* rather than man-made creations like houses or songs.
Is mathematics invented or discovered? The Greeks came down
emphatically in favor of "discovered," but the question is ancient,
like *what is justice?* and apparently just as difficult to resolve.

On the one hand, what could more plainly be human inventions
than the concepts of geometry and algebra? Even the simplest
mathematical notion has no tangible existence in the everyday
world. Who goes for a walk and trips over a 3? On the other

hand, what could be more obvious than that the truths of mathematics are facts about the world, whether or not any humans catch on to them? If two dinosaurs were at a watering hole and two more dinosaurs came along to join them, the science writer Martin Gardner once asked, weren't there four dinosaurs altogether? Didn't three stars in the sky form a triangle before the first humans came along to define triangles?*

Newton and the other scientists of the seventeenth century shared the Greek view, and they coupled it with their own fundamental belief that the world was a cosmic code, a riddle designed by God. Their mission, in the words of one prominent writer of the day, was to decode that "strange Cryptography." The Greeks had harbored similar ambitions, but the new scientists had advantages their predecessors had lacked. First, they had no taboos about studying motion mathematically. Second, they had calculus, a gleaming new weapon in the mathematical arsenal, to study it with.

Just as important, they had complete, unbreakable faith that the riddle had an answer. That was vital. No one would stick with a crossword puzzle if they feared that the areas of order might be mixed with patches of gibberish. Nature presented a greater challenge than any crossword, and only the certain knowledge that God had played fair kept scientists struggling year after year to catch on to His game.

---

* The nineteenth-century German mathematician Carl Gauss, a towering figure in the history of mathematics, believed in the possibility of life on other worlds. Gauss supposedly proposed—the story may well be apocryphal—that since all intelligent beings would eventually discover the same mathematical truths, we could communicate with moon creatures by choosing a vast, empty space in Siberia and planting trees in an enormous diagram of the Pythagorean theorem.

Even so, the task was enormously difficult. Nonmathematicians underestimated the challenge. When Francis Bacon spoke of the mysteries of science, for instance, he made it sound as if God had set up an Easter egg hunt to entertain a pack of toddlers. God "took delight to hide his works, to the end to have them found out."

Why would God operate in such a roundabout way? If his intent was to proclaim His majesty, why not arrange the stars to spell out BEHOLD in blazing letters? To seventeenth-century thinkers, this was no mystery. God *could* have put on a display of cosmic fireworks, but that would have been to win us over by shock and fear. When it came to intellectual questions, coercion was the wrong tool. Having created human beings and endowed us with the power of reason, God surely meant for us to exercise our gifts.

The mission of science was to honor God, and the best way to pay Him homage was to discover and proclaim the perfection of His plans.

Chapter Twenty-Four

THE SECRET PLAN

When Newton declared that he stood on the shoulders of giants, he was at least partly sincere. He did genuinely admire some of his fellow scientists, particularly those who'd had the good judgment to die before he came along. One of the great predecessors he had in mind was the astronomer Johannes Kepler. A contemporary of Galileo, Kepler was a genius and a mystic whose faith in God and faith in mathematics had fused into an inseparable unit.

Kepler was both astronomer and astrologer, though he never sorted out just how much the heavens influenced human affairs. "In what manner does the countenance of the sky at the moment of a man's birth determine his character?" he wrote once, and then he answered his own question. "It acts on the person during his life in the manner of the loops which a peasant ties at random around the pumpkins in his field: they do not cause the pumpkin to grow, but they determine its shape. The same applies to the sky: it does not endow man with his habits, history, happiness, children, riches or a wife, but it molds his condition."

For many years the sky seemed set against Kepler. He grew up poor, sick, and lonely. His childhood, according to an account he compiled later, was a long series of afflictions ("I was born premature . . . I almost died of smallpox . . . I suffered continually from skin ailments, often severe sores, often from the scabs of chronic putrid wounds in my feet"). He remained adrift into his twenties, cut off from others not only by his intelligence but also by his quarrelsome, touchy, defensive manner. "That man has in every way a dog-like nature," he wrote, for some reason describing himself in the third person. "His appearance is that of a little lap-dog. . . . He liked gnawing bones and dry crusts of bread, and was so greedy that whatever his eyes chanced on he grabbed."

Kepler was brilliant but restless, hopping from obsession to obsession. Astrology, astronomy, theology, mathematics all captivated him. They related to each other in some way that he could sense but not articulate. After his own university days, he managed to find work as a high school teacher, but his students found him disorganized and hard to follow, and soon his classroom was nearly deserted. And then, on a summer day, while teaching a class on astronomy, Kepler had his *Eureka!* moment. To the end of his life, he would remember the instant when he glimpsed God's blueprint.

It was July 9, 1595. Kepler was twenty-four years old, and he believed fervently in Copernicus's doctrine of a sun-centered universe. For weeks he had been laboring to find some pattern in the planets' orbits. If you knew the size of one planet's orbit, what did that tell you about the others? There had to be a rule. Kepler tried ever more complicated numerical manipulations. Each one failed. Now, standing at the front of his classroom, he began drawing a diagram having to do with the positions of Jupiter and Saturn, the two most distant planets then known. Kepler knew the size of both orbits, but he couldn't see any connection between the two.

Jupiter and Saturn were important astrologically—our words *jovial* and *saturnine* are fossils of bygone doctrines—and what was especially important were the times the two planets were "in conjunction," near one another in the sky. If they met at a certain point today, astronomers knew, they would meet next (in twenty years) at a point 117 degrees away, just under one-third of the way around the zodiac. The conjunction point after that one would be another 117 degrees along, and so on. Kepler drew a circle showing the first conjunction point, the second, and the third.

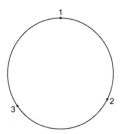

The diagram shows where Saturn and Jupiter appear in the sky together. If today they can be seen at point 1, in twenty years they will appear at 2, in twenty more at 3, and so on.

He filled in more conjunction points, each one 117 degrees along from its predecessor. (If the points had been 120 degrees apart, exactly one-third of the way around a circle, there would have been a total of only three conjunction points, because all the points after the first three would have overlapped.)

Continuing in the same way, Kepler soon had a circle with evenly spaced, numbered dots marked all the way around it. (Look at the diagram below, in which points 1 through 5 are labeled.) Each dot represented a point where Saturn and Jupiter met.

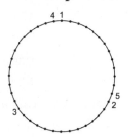

For no especially clear reason, Kepler drew a line from the first conjunction to the second, from the second to the third, and on and on. From that series of straight lines emerged, mysteriously and unexpectedly, not some straight-sided shape but a new circle. To Kepler it seemed as if his original circle had conjured up a new, smaller counterpart inside itself.

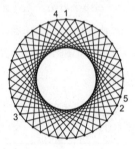

Staring at that circle within a circle, Kepler found himself almost staggering. (Kepler would have loved *The Da Vinci Code*.) At the moment that the new, inner circle swam into his view, he saw the secret plan behind the universe's design. "The delight that I took in my discovery," he wrote, "I shall never be able to describe in words."

Only a skilled geometer with a bone-deep faith that God himself delighted in geometric riddles would have seen anything noteworthy in Kepler's drawing. But Kepler, who knew that nothing in nature is mere coincidence, looked at his two circles and thought of his two planets, and marveled. What could it mean except that the outer circle represented the orbit of the outmost planet, Saturn, and the inner circle the orbit of the inner planet, Jupiter? And the inner circle was half the size of the outer circle, just as Jupiter's orbit was half the size of Saturn's!

But that was only the start. Kepler's full discovery had an even more mystical, more geometric flavor. Saturn and Jupiter were the first two planets (counting from farthest from the sun to nearest). What connected their orbits? What else was "first"?

The answer struck Kepler like a hammerblow. *This* was the eureka insight. "The triangle is the first figure in geometry," Kepler exclaimed—"first" in this case meaning "simplest"—and that first, simplest geometric figure was the key to the mystery of the first two orbits. Kepler had known all along that Saturn's orbit and Jupiter's orbit could be depicted as a circle inside a circle, but there are countless ways to draw one circle inside another. The mystery Kepler yearned to solve was why God had chosen *these* two circles in particular. The triangle gave him the answer.

Feverishly, Kepler put his brainstorm to the test. He drew a circle and inside it he drew the one triangle that stood out from all the other possibilities—the simplest triangle of all, the only one that fit perfectly inside the circle and was completely symmetrical, with all three sides identical. Inside the triangle he drew another circle. Again, he could have chosen any of countless circles; again, he made the only "natural" choice, the one circle that fit the triangle perfectly. He looked again at his drawing. The inner circle nestled snugly inside the triangle, and the triangle fit neatly and naturally into the outer circle. In Kepler's mind, the outer circle represented Saturn's orbit, the inner circle Jupiter's. The triangle that tied the two together was the first shape in geometry. Kepler stared at that geometric emblem.

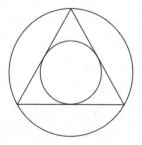

He performed a quick calculation—the outer circle in his diagram was twice the circumference of the inner circle. And Saturn's orbit was twice Jupiter's. He had broken God's code. Now Kepler set to work in a frenzy. If the orbits of the first two planets depended on the simplest geometric shape, a triangle, then the orbits of the *next* two planets must depend on the next simplest shape, a square.

Kepler drew a circle, representing Jupiter's orbit. The question was what circle would represent the orbit of the next planet toward the sun, Mars. In Kepler's mind, the answer nearly shouted aloud. Inside the Jupiter circle, he drew a square. Inside that square he drew the one, special, God-designated circle that fit perfectly. That inner circle depicted Mars's orbit.

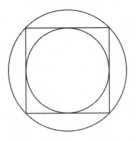

And Kepler could continue in this way for all the planets, working his way in toward the sun, arraying the planets just as Copernicus had shown them. The orbits were nested one within the other, and the size of one automatically dictated the size of the next. The first two orbits were built around a triangle, which has three sides; the next two around a square, with four sides; the next two around a pentagon, with five sides; and so on. Kepler set to work drawing squares, pentagons, hexagons, septagons, with circles in between.

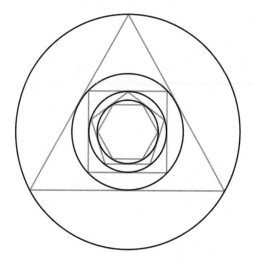

Kepler believed that God had arranged the planets' orbits according to this geometric scheme. (For clarity, the diagram shows only the four outermost planets, not all six planets known in Kepler's day.)

Johannes Kepler had discovered the architecture of the solar system. Or so he believed, and in his fever dream he filled sheet after sheet with ever more elaborate geometric diagrams. For the young, unknown astronomer, this was dizzyingly exciting. Without looking out the window, he had not only assigned each planet to its proper place but shown why it *had* to occupy that place.

It was perfect, it was elegant, and it was wrong. As Kepler took more time to compare the actual sizes of the planets' orbits with the sizes his model predicted, he found mismatches he couldn't explain away. He tried endless fixes. Nothing. How could God have led him astray?

Chapter Twenty-Five

TEARS OF JOY

At last the light dawned. Kepler had been thinking in two dimensions, in the flat world of circles and triangles and squares. But the universe has three dimensions. How much time had he wasted? "And now I pressed forward again. Why look for two-dimensional forms to fit orbits in space? One has to look for three-dimensional forms—and, behold dear reader, now you have my discovery in your hands!"

The switch to three dimensions represented far more than a chance to salvage a pet theory. One of the riddles that tormented Kepler had to do with the number of planets—there were exactly six. (Uranus, Neptune, and Pluto were not yet known.*) Why had God picked six, Kepler asked, "instead of twenty or one hundred"? He had no idea, and all his fussing with squares and pentagons and hexagons had brought him no nearer to an answer.

But now he realized that he had overlooked a glaring clue. Euclid had proved, two thousand years before, that in three dimensions the story of symmetrical shapes has an extraordinary

---

* Pluto is considerably smaller than the moon, and in 2006 astronomers decided to downgrade it to "minor planet" status.

twist. Working in two dimensions, you can draw an endless succession of perfectly symmetrical, many-sided figures—triangles, squares, pentagons, hexagons, and so on, forever. If you had enough patience, you could draw a hundred-sided polygon or a thousand-sided one. (All you'd have to do is draw a circle, mark equally spaced dots on it, and then connect each one to its next-door neighbors.)

In three dimensions, where there is more room, you might expect the same story—a handful of simple shapes like pyramids and cubes and then a cascade of increasingly complicated ones. Just as a pyramid is made of triangles pasted together, and a cube is made of squares, so you might guess that you could glue fifty-sided shapes together, or thousand-sided ones, and make infinitely many new objects.

But you can't. Euclid proved that there are exactly five "Platonic solids"—three-dimensional objects where each face is symmetrical and all the faces are identical. (If you needed dice to play a game, the mathematician Marcus du Sautoy points out, these five shapes are the only possible ones.) Here is the complete array. There are no others:

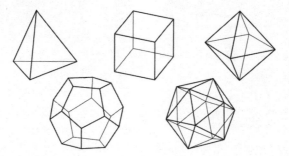

Only five. And there are six planets. *Now* Kepler had it. He still had to work out the details, but at last he'd seen the big picture. Each planet traveled around the sun, its orbit confined to a particular sphere. The spheres sat one inside the other. But what determined the sizes of the spheres? God, the greatest of all geometers, surely had a plan. After a false start, Kepler had seen it. Each sphere fit snugly and symmetrically inside a Platonic solid. Each Platonic solid, in turn, fit snugly and symmetrically inside a larger sphere. In a flash, Kepler saw why God had designed the cosmos to have six planets and why those orbits have the sizes they do. He burst into tears of joy.

"Now I no longer regretted the lost time," he cried. "I no longer tired of my work; I shied from no computation, however difficult." On and on he calculated, computing orbits, contemplating octahedrons and dodecahedrons, working without rest in the hope that at last he had it right but always terrified that once again his "joy would be carried away by the winds."

But it wasn't. "Within a few days everything fell into its place. I saw one symmetrical solid after the other fit in so precisely be-

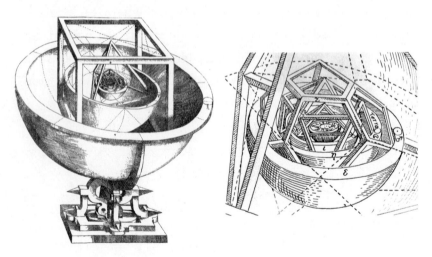

Kepler devised a new, more elaborate scheme to explain the planets' orbits. God had built the solar system around the five "Platonic solids." The diagram at right, with the sun at its center, is a detail of the drawing at left. The sun sits inside a nested cage; the inmost shape is an octahedron.

tween the appropriate orbits, that if a peasant were to ask you on what kind of hook the heavens are fastened so that they don't fall down, it will be easy for you to answer him."

*   *   *

Kepler rejoiced in his success. "For a long time I wanted to become a theologian," he told an old mentor. "For a long time I was restless. Now, however, behold how through my effort God is being celebrated through astronomy."

In 1596 he presented his theory to the world in a book called *The Mystery of the Universe*. Even with his book completed, Kepler fretted about whether his model fit the actual data about the planets' orbits quite well enough. For the time being, he managed to fight down his doubts. He happily devoted long hours to constructing models of his solar system from colored paper and drawing plans for a version made of silver and adorned with diamonds and pearls. "No one," he boasted, "ever produced a first work more deserving of admiration, more auspicious and, as far as its subject is concerned, more worthy."

In the decades to come Kepler would make colossal discoveries, but his pride in his elaborate geometric model never faded. Centuries later the biologist James Watson would proclaim his double helix model of DNA "too pretty not to be true." Kepler had felt the same joy and the same certainty, but eventually the data left him no choice but to acknowledge that he had gone wrong, again.

His perfect theory was only a fantasy, but it proved enormously fruitful even so. For one thing, *Mystery of the Universe* transformed Kepler's career. He sent a copy of the book to Tycho Brahe, the leading astronomer of the day, who found it impressive. In time Kepler would gain access to Tycho's immense and meticulous trove of astronomical data. He would pore over those figures incessantly, over the course of decades, trying to make his model work and uncovering other patterns concealed in the night sky. Later scientists would rummage through Kepler's collection of numerical discoveries and find genuine treasure among the dross.

Kepler valued *Mystery of the Universe* so highly because it was there that he had unveiled his great breakthrough. But in the course of discussing his model of the heavens, he had scored another history-making coup. Kepler had followed Copernicus in placing the sun at the center of his model, but then Kepler had moved a crucial step beyond all his predecessors. Not only did all the planets circle the sun, he noted, but the farther a planet was from the sun, the slower it traveled in its orbit. Somehow the sun must *propel* the planets, and whatever force it employed plainly grew weaker with distance.

Kepler had not yet found the law that described that force— that would take him another seventeen grueling years—but this was a breakthrough even so. Astrologers and astronomers had always focused their attention on mapping the stars and charting the planets' journeys across the sky. The goal had been description and prediction, not explanation. No one before Kepler had ever focused on asking what it was that moved the planets on their way. From now on, scientists looking at the heavens would picture the stars and planets as actual, physical objects pushed and tugged by some cosmic engine and not simply as dots on a chart.

"Never in history," marvels the historian of science Owen Gingerich, "has a book so wrong been so seminal in directing the future course of science."

Chapter Twenty-Six

WALRUS WITH A GOLDEN NOSE

From the start Kepler's faith that God was a mathematician both impeded him and spurred him. First his faith lured him into devoting years to his Platonic pipe dream; when that dream dissolved it motivated him to search elsewhere, in the certain knowledge that there had to be *some* mathematical pattern that explained the solar system. Through all his years of searching, Kepler's fascination was less with the objects in the sky—the sun, stars, and planets—than with the relationships among them. Not the things but the patterns. "Would that God deliver me from astronomy," Kepler once wrote, "so I can devote all my time to my work on harmonies."

In time that work would yield many patterns, a few of them among the highest achievements of human thought but most of them nearly inexplicable to modern readers. When Kepler finally abandoned his elaborate geometric model of the planets, for instance, he replaced it with an equally arcane model based on music. This new search for "harmonies" built on Pythagoras's age-old insight about strings of different lengths producing notes of different pitch. Kepler's notion was that the planets in their various orbits, traveling at different speeds, corresponded to different musical notes, and "the heavenly motions are nothing but a continuous song for several voices (perceived by the intellect, not by the ear)."*

Kepler's new system, with its sopranos and tenors and basses, was as farfetched as its predecessor, with its cubes and pyramids and dodecahedrons. As it turned out, neither model had anything to do with reality. But in the course of his obsessive, misguided quest to prove the truth of his theories, Kepler did make genuine, epochal discoveries. Scientists would eventually dub three of these "Kepler's laws," though Kepler never gave them that name nor deemed them any more praiseworthy than his other finds.†

Late in his life, when he looked back over his career, Kepler himself could scarcely pick out his breakthroughs from the mathematical fantasies that surrounded them. "My brain gets tired when I try to understand what I wrote," he said later, "and I find it hard to rediscover the connection between the figures and the text, that I established myself."

Kepler was one of the most daring, insightful thinkers who ever lived, but his career only took off when he joined forces with an astronomer who was his opposite in almost every respect. Kepler was poor and lean, a creature of ribs and patches. Tycho Brahe was rich beyond measure. Kepler was shy and ascetic, Tycho

---

* Not by the human ear, at any rate. God could hear these cosmic harmonies, as dogs can detect whistles pitched too high for human hearing.
† The first person to refer to Kepler's "laws" was Voltaire, in 1738. Scientists eventually followed his lead.

hard-drinking and rowdy. Kepler was imaginative and creative, sometimes alarmingly so, Tycho a brilliant observer but a run-of-the-mill theoretician. But the two great astronomers needed one another.

Tycho* was a Danish nobleman with a private observatory on a private island. The most eminent astronomer of the generation before Kepler and Galileo, it was Tycho who had startled the world in 1572 by proving that the new star that had somehow materialized in the sky truly was a star. Nothing about Tycho was run-of-the-mill. Round, bald, sumptuously dressed, he looked like Humpty Dumpty with a walrus mustache and a velvet cloak. He ruled his mini-kingdom like a mini-king, presiding over lavish banquets and cackling over the antics of his court jester, a dwarf named Jepp.

In his student days, Tycho had lost part of his nose in a sword-fight. In one version of the story, the trouble started at a wedding celebration when another wealthy young Dane reminded everyone of some odd events from a few months before. Tycho had announced with great fanfare, in a poem written in elegant Latin flourishes, that a recent eclipse of the moon foretold the death of the Turkish sultan. But, it turned out, the sultan had died six months *before* the eclipse. Tycho's rival told the story with gusto, and nearly all his listeners enjoyed it. Not Tycho. The retelling of the story led to bad blood and, soon after, a duel. Tycho nearly lost his life and did lose a chunk of his nose. For the rest of his life he sported a replacement made of gold and silver.

Despite the bluster and showmanship, Tycho was a genuine scholar. His observatory was the best in Europe, outfitted with a dazzling array of precision-made sextants, quadrants, and other

---

* Tycho, like Galileo, is generally referred to by his first name.

devices for pinpointing the positions of stars. The observatory stood in a grand, turreted castle that boasted fourteen fireplaces and an astonishing luxury, running water. In Tycho's library stood a celestial globe five feet in diameter and made of brass; when a star's position was established beyond a doubt, a new dot was carefully added to the globe. Tycho boasted that his observatory had cost a ton of gold, and Kepler complained that "any single instrument cost more than my and my whole family's fortune put together."

Kepler had sent his *Mystery of the Universe* to Tycho and all the other eminent scientists he could think of. Many could not fathom what he was up to. Tycho, more mystically minded than Galileo and some of the other skeptics, replied enthusiastically and soon took Kepler on as his assistant. It was an arrangement with obvious benefits for both men. Tycho had devised a hybrid model of the solar system, partway between the ancient Earth-centered model and Copernicus' sun-centered version. In this picture, the sun and moon orbited the Earth, and the five other planets orbited the sun. Tycho had compiled reams of scrupulously accurate observations, but without Kepler's mathematical help he could not demonstrate the truth of his hybrid model. Kepler had no interest in Tycho's model, but in order to make progress on his own theories he desperately needed Tycho's records.

But Tycho hoarded them. Torn between hope that the younger man could find patterns hidden within twenty years' worth of figures and fear that he was giving away his treasure, Tycho clung to his numbers with a miser's grip. Kepler snarled helplessly. And then, out of the blue, Tycho died. (He died of a bladder infection brought on, according to Kepler, by drinking too much at a banquet and refusing to leave the table to pee.) Kepler had been with Tycho only eighteen months, but now he had what he needed. "I was in possession of the observations," Kepler noted contentedly, "and refused to hand them over to the heirs."

Chapter Twenty-Seven

CRACKING THE COSMIC SAFE

For nearly twenty years Kepler would stare at Tycho's figures, certain that they concealed a hidden message but for months and years at a time unable to make any headway in deciphering it. He had long since abandoned his geometric models, on the grounds that they simply did not fit the data. The problem was that nothing else did, either.

Kepler knew, for example, how long it took each planet to orbit the sun—Mercury, 3 months; Venus, 7 months; Earth, 1 year; Mars, 2 years; Jupiter, 12 years; Saturn, 30 years—but try as he might he could not find a rule to connect those numbers. This was a task with some resemblance to making sense of the numbers on a football scoreboard if you had never heard of football. The number 3 appears sometimes and so do 7 and 14, but never 4 or 5. What could be going on?

Even armed with Tycho's astronomical data, Kepler took six years to find the first two of the three laws now named for him. The story of Kepler's discovery of his laws is a saga of false starts and dead ends piled excruciatingly one upon another, while poor Kepler despaired of ever finding his way.

Kepler's first law has to do with the paths the planets travel as they orbit the sun. Kepler shocked his fellow astronomers—he shocked himself—by banishing astronomy's ancient emblem of perfection, the circle. But Tycho's data were twice as accurate as any that had been known before him, and Kepler, who had indulged himself in endless speculative daydreams, now turned the world upside down because of a barely discernible difference between theory and reality. "For us, who by divine kindness were given an accurate observer such as Tycho Brahe," Kepler wrote, "for us it is fitting that we should acknowledge this divine gift and put it to use." To take Tycho's measurements seriously meant to acknowledge, albeit slowly and reluctantly, that the planets simply did not travel in circles (or in circles attached to circles or any such variant).

Worn down by endless gruesome calculations, Kepler nearly despaired of ever finding the patterns hidden inside the astronomical records. (He referred wearily to his hundreds of pages of calculations as his "warfare" with the unyielding data). Finally he found that each planet orbits the sun not in a circle but in an ellipse, a kind of squeezed-in circle. This meant, among other things, that the distance from the sun to a planet was not constant, as it would be if the planet traveled in a circle, but rather always changing.

All circles are identical except in size—this was part of what made them perfect—but ellipses come in infinite variety, some barely distinguishable from circles and others long and skinny. An ellipse is not just an oval but an oval of a specific sort. (To draw an ellipse push two tacks into a piece of cardboard and drop a loop of string around them. Pull the string taut with a pencil and move the pencil along. Each tack is called a focus. The defining property of an ellipse is that, for every point on the curve, if you measure the distance from one focus to the pencil tip and then add to that number the distance from the pencil tip to the other focus, the sum is always the same.*)

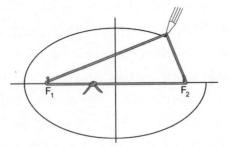

For every point on the ellipse, the distance from F$_1$ to the pencil's tip plus the distance from F$_2$ to the pencil tip is the same.

In the case of the planets, Kepler found, the sun sits at one focus of an ellipse. (The other focus does not correspond to a physical object.) This was Kepler's first law—the planets travel in an ellipse with the sun at one focus. This was truly radical. Even Galileo, revolutionary though he was, never abandoned the belief that the planets move in circles.

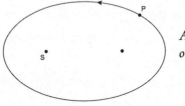

According to Kepler's first law, the planets orbit the sun not in a circle but in an ellipse.

* * *

Kepler's second law was heretical, too. It had to do with the planets' speed as they travel, and it involved another assault on

---

* A circle can be thought of as a special ellipse, one in which the two focuses are in the same place.

uniformity. The planets didn't travel in perfect circles, Kepler claimed, and they didn't travel at a steady pace, either. The spur was Kepler's belief that the sun somehow pushed the planets on their way. If so, it stood to reason that the force pushed harder when a planet was near the sun and more weakly when it was farther away. When a planet neared the sun, it would race along; when far away, it would dawdle.

It took Kepler two years of false starts to find his second law. (He earned his living, in the meantime, as imperial mathematician to Rudolph II, the Habsburg emperor whose court was in Prague. Kepler's official duties largely centered on such tasks as preparing horoscopes and making astrology-based forecasts of next season's weather or a stalemated war's outcome.) His great insight was finding a way to capture the planets' uneven motion in a precise, quantitative rule. The natural way to describe a planet's motion was to chart its position every ten days, say, and then compute the distance between one point and the next. But that procedure turned out not to reveal any general rule. In a moment of inspiration Kepler saw a better way. The key was to

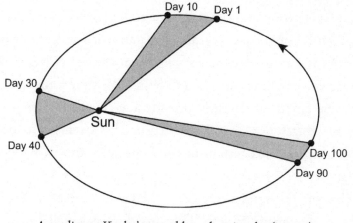

According to Kepler's second law, the triangles (in gray)
have different shapes but are identical in area.

think not of distance, which seemed natural, but of area, which seemed irrelevant.

Kepler's second law: the line from a planet to the sun sweeps out equal areas in equal times.

Proud of his discoveries though he was, Kepler had no great fondness for these laws, because he had no idea where they'd come from. Why had God not employed circles? Circles were perfect; ovals and ellipses were, Kepler lamented, a "cartload of dung." And if for some reason He had chosen ovals, why choose ellipses in particular rather than egg shapes or a thousand other possibilities?

Kepler's third law seemed the most arbitary of all, and proved the hardest to find. Kepler's first two laws had to do with the planets considered one at a time. His third law dealt with all the planets at once. Kepler's quest was once again, just as it had been with his Platonic solids model, to find what the various orbits had to do with one another. God surely had not set the planets in the sky arbitrarily. What was His plan?

Kepler had two sets of numbers to work with—the size of each planet's orbit and the length of each planet's "year." Neither set of figures on its own revealed any pattern. The size of Earth's orbit, for instance, revealed nothing about the size of Mars's orbit, nor did the length of one planet's "year" (the time it took to complete one circuit of the sun) provide any clue to the length of a year on a different planet. Kepler turned his attention to looking at both numbers together, in the hope of finding a magic formula.

The general trend was clear—the farther a planet was from the sun, the longer its year. That stood to reason, because near-to-the-sun planets had small orbits to trace and distant planets big ones. But it wasn't a matter of a simple proportion. Planets farther from the sun had more distance to cover than closer-in planets *and* they traveled more slowly. It would be as if ships crossing the ocean traveled more slowly than ships hopping along the coast from port to nearby port.

Since he had no idea of the forces that moved the planets, Kepler took on the code-breaking challenge as if it were purely a task in numerology. Like a safecracker armed with nothing but patience, he tried every combination he could think of. If there was no pattern in the lengths of the different planets' years, for instance, perhaps there was a pattern if you took the lengths of the years and squared them. Or cubed them. Or perhaps you could see a pattern if you computed each planet's maximum speed and compared those. Or the minimum speeds. For more than a dozen years, Kepler tried one combination after another. He failed every time.

Then, out of the blue, "On March 8 of this present year 1618, if precise dates are wanted, the solution turned up in my head." The discovery itself was complicated. Characteristically, so was Kepler's response, which combined gratitude to God, immense pride in his own achievement, and his customary willingness to paint himself unflatteringly. "I have consummated the work to which I pledged myself, using all the abilities that You gave to me; I have shown the glory of Your works to men," he wrote, "but if I have pursued my own glory among men while engaged in a work intended for Your glory, be merciful, be compassionate, and forgive."

What Kepler had found was a way—a mysterious, complicated way—to tie the orbits of the various planets together. It required that you perform a messy calculation. Choose a planet, Kepler said, and then take its orbit and cube it (multiply it by itself three times). Next, take the planet's year and square it (multiply it by itself). Divide the first answer by the second answer. For every planet, the result of that calculation will be the same. Kepler's third law is the assertion that if you follow that unappetizing recipe the answer always comes out the same.

Kepler knew, for instance, that Mars's distance from the sun is 1.53 times Earth's distance, and Mars's year is 1.88 times Earth's year. He saw—somehow—that $1.53 \times 1.53 \times 1.53 = 1.88 \times 1.88$. The other planets all told the same story. (Put another way, the length of a planet's year depends not on its distance from the sun, or on that distance squared, but on something in between—the distance raised to the $3/2$ power.)

But why? What did it mean?

The numbers worked out, which seemed beyond coincidence, but it all sounded like mumbo jumbo. Of all the ways that God might have arrayed the planets and set them orbiting on their way, why had He picked one built around this curious business of squaring and cubing?

The safe door had swung open, but Kepler had no idea why.

Chapter Twenty-Eight

THE VIEW FROM
THE CROW'S NEST

Kepler's laws represented a giant advance in decoding God's cryptography, even if he did not know why they were true or what they meant. The next advance came from his fellow astronomer, Galileo, who was almost exactly Kepler's contemporary.

Galileo was born in the same year as Shakespeare, and Galileo's stature in science nearly matches Shakespeare's in literature. "I believe that if a hundred of the men of the seventeenth century had been killed in infancy, the modern world would not exist," wrote Bertrand Russell. "And of these hundred, Galileo is the chief." In truth, that seems unlikely. Galileo's genius is beyond dispute, but every great scientist, from Galileo to Darwin to Einstein, had rivals at his heels. If Shakespeare had not lived, we would not have "to be or not to be." If Einstein had not lived, we might have had to wait a few years for $e = mc^2$.

The same holds for Galileo, great as he was. But Galileo did shake the scientific world out of its doldrums, and perhaps Russell was right that no one could have been better suited to that task, either temperamentally or intellectually. Galileo was brilliant, cantankerous, and expert in wielding all the weapons of intellectual combat. (Even his hair bristled, as if it, too, were poised for battle.) He spoke wittily and wrote vividly, with a knack for metaphors and homey analogies; he had a flair for mockery, name-calling, and sarcastic put-downs; he had, when he wanted, a honeyed tongue and, as an acquaintance noted, "a way of bewitching people."

Curiously, Galileo did *not* build on Kepler's work. Indeed, he seems not to have known that Kepler's laws existed at all, even though Kepler had sent him his *New Astronomy*, which included the first two laws and endless astronomical speculation besides. (Galileo put the book away unread.) What Galileo did instead was turn to a completely different part of the cosmic riddle.

Galileo focused on solving a mystery far older than Kepler's. Nearly a century after Copernicus, the question *does the Earth move?* still struck nearly everyone as absurd. Armed with the fame he had earned by turning his telescope to the heavens, Galileo set out to do what Copernicus and Kepler had never done, and what Kepler's laws did not do, either—find a response to the claim that the Earth could not possibly be in motion. Then he set out to tell the world.

Galileo wrote his most important scientific works not in formal, impenetrable prose but in pugnacious dialogues, like miniature plays. He put the arguments of his rivals into the mouth of a character he named Simplicio, the embodiment of intellectual mediocrity. The name perhaps referred to an actual figure, an Aristotelian named Simplicius who lived around a thousand years before Galileo. More likely, Galileo made up "Simplicio" because it was so close to *simpleton* (*sempliciotto* in Italian). Certainly his readers jumped to that conclusion. And one of them, Pope Urban VIII, was notably unamused to find a pet argument of his own put in Simplicio's mouth.

That proved to be a disastrous miscalculation, but a telltale one. Galileo's Italy was a flamboyant place. Showmanship was far more common than bashfulness, and Galileo was never much inclined to hide his talent in any case. Still, he dangerously overestimated his own powers of persuasion. He liked holding forth, about wine and cheese and literature to be sure, but especially about the excellence of the new picture of the heavens and the foolishness of the old view. "He discourses often amid fifteen or twenty guests who make hot assaults upon him, now in one house, now in another," one friend recalled, after a dinner party, "but he is so well buttressed that he laughs them off."

To be outnumbered was part of the fun. "If reasoning were like hauling," Galileo proclaimed, "I should agree that several reasoners would be worth more than one, just as several horses can haul more sacks of grain than one can. But reasoning is like racing and not like hauling, and a single Arabian steed can outrun a hundred plowhorses."

Galileo not only defended Copernicus against his critics but, in the course of making his argument, devised a theory of relativity. Three centuries before Einstein's version, Galileo's theory proved nearly as hard for common sense to grasp. Inside a room with the curtains pulled, Galileo showed, there is no way to tell if you're standing still or traveling in a straight line at a steady speed. Inside the compartment of a smooth-running new train, to take a modern example, no experiment you can do (short of peeking out the window) will reveal whether you're sitting motionless or racing down the track. You might think that dropping your keys would give the game away—if the train is moving to the east, wouldn't the keys fall a bit toward the west?—but in fact they fall straight down, as usual.

More to the point, what is true of a ship or a train is true of the Earth itself—there is no way to tell if the Earth is moving or standing still, short of carrying out sophisticated astronomical measurements. No ordinary actions we can carry out reveal whether we're moving. The same holds true for *any* motion that is smooth, steady, and straight, no matter how fast it might be. (The Earth's orbit is nearly circular, not straight, but the circle is so huge, compared to our speed, that any short stretch is effectively a straight line.)

This was a direct attack on Aristotle and all his followers. We can be sure that the Earth does not move, Aristotle had insisted, because we see proof everywhere we look. Rocks fall straight down, not on some curved or slanted path. Buildings don't shake or topple, as they would if the ground beneath them was on the move. A moving world would be chaotic, Aristotle taught, and the most routine task would be as difficult as trying to paint a room while standing on a ladder mounted on wheels.

Galileo showed this was false. Nothing is special about a motionless world. Smooth, steady motion looks and feels exactly the same as utter stillness. The strongest argument against Copernicus—that he began by assuming something that was plainly ridiculous—was invalid.

Galileo reached these far-ranging conclusions by means of the humblest experiments imaginable. He began with a metal ball and a wooden ramp. (In time he would add a bucket of water with a hole poked in it.)

Galileo's pet subject was motion, in particular the motion of falling objects. For Aristotle, as we have seen, to be in motion meant to change—from one position to another, perhaps, but also from one "quality" to another, as from "foolishness" to "wisdom." Galileo was after what looked like simpler game. He wanted to know the rules that govern inanimate objects in free fall. But how could he look closely enough to tell precisely how rocks plummet?

The answer, he decided, was to slow things down. Rather than drop a rock through the air, he would roll a ball down a ramp and hope that what held for the ramp would hold for free fall. This was a nervy move. Arguments by analogy are always risky, and here the analogy seemed far from ironclad. But Galileo, a brilliant teacher and debater, presented this leap of faith as if it were but another step on a casual walk, and his audience leaped with him.

He began by seeing what happened when he let a ball roll down one ramp, across a table, and then up a second ramp. If the two ramps were identical, it turned out, the ball ended up at virtually the same height it had started at. (In the same way, if you let go of a marble in a circular bowl it will roll to the bottom and then up the other side to very, very near its starting height.)

Then came the crucial observation. Galileo chose a second ramp that was less steep than the first one. Once again the ball ended up at the height it had started at, though this time it had to roll farther to get there. Then still another repetition, this time with a second ramp that was only tilted ever so slightly. Again, the ball eventually reached its starting height but it had to roll and roll to get there.

And suppose the second ramp was perfectly flat, not tilted at all? Then, said Galileo, the ball would roll horizontally *forever.* The flat ramp was a thought experiment, not a real one, but Galileo proclaimed a new law of nature—any object moving horizontally will continue moving horizontally forever, at the same speed, unless something comes along to intervene. (Newton's first law of motion is a generalization of the same principle.) Aristotle had decreed exactly the opposite, as we saw earlier. In Aristotle's world, motion was unnatural and always called for explanation; unless a force kept pushing or pulling it, a moving object would always slow and then stop.

We should not downplay Galileo's boldness. In rejecting Aristotle, he was also dismissing what everyone has seen for themselves countless times—moving objects *do* always stop. Ignore what all your experience and your common sense have taught you, Galileo said. More important than the world you actually see, more true to the essential nature of things, is an idealized, abstract, mathematical world that you can only see with the mind's eye.

In Galileo's hands, the simple statement that motion was natural had enormous consequences. Here was the key to his theory of relativity and the rebuttal to the Aristotelians' guffawing about a moving Earth. In his day, when roads were rutted and coaches horse-drawn, the most familiar example of smooth travel was on shipboard. What would happen, Galileo asked, if a sailor climbed to the top of the mast and dropped a rock? For Aristotle, this would have been an easy question. If the ship was motionless, at rest in a quiet harbor, the rock would fall straight down and hit the deck at the base of the mast. If the ship was gliding along on a glassy sea, the rock would crash to the deck several inches from the mast. Galileo disagreed. In *both* cases, he insisted, the rock would fall straight down to the base of the mast.

The reason was his first law. The ship, the sailors, the passengers, the rock falling from the mast, are all in horizontal motion, all of them moving together. The rock lands at the base of the mast because mast and rock are both moving horizontally, in unison, at the same time as the rock is hurtling downward.

"Shut yourself up with some friend in the main cabin below decks on some large ship," Galileo wrote. Bring in some butterflies, a fishbowl with some fish swimming around, a leaky jug dripping water into a pan on the floor. No matter how closely you looked for something out of the ordinary (the fish clustered against one side of their bowl, for instance, or the drops of water missing the pan), Galileo went on, "you could not tell from any of them whether the ship was moving or standing still."

The same holds for the Earth itself, and all its passengers, as it speeds along on its voyage. The speeding Earth, which races in its orbit at about eighteen miles *per second*, is as safe and solid a home as a ship safely moored at anchor in a mirror-smooth sea.

Even today, Galileo's insight doesn't come naturally. We believe him, though, because we've all carried out countless tests of our own. It sometimes happens, for instance, that we're speeding down the highway, with the car windows rolled up, when we notice a fly buzzing around. The car might be traveling at 70 miles per hour, much faster than any fly can manage, and yet the fly continues unperturbed. Why doesn't the back window slam into it at 70 miles per hour?

Or think about traveling by plane. Drop your phone in a jet and during the fraction of a second it takes to hit the floor, the plane will have traveled perhaps a hundred yards. How is it that it falls at your feet and not a football field behind you? For that matter, how can the flight attendants dare to pour coffee? While the coffee is in midair, on its way to the cup but not yet there, the cup itself will have moved hundreds of feet. How can the crew serve first-class without scalding everyone in economy?

"A company of chessmen standing on the same squares of the

chessboard where we left them, we say are all in the same place or unmoved: though perhaps the chessboard has been in the meantime carried out of one room into another." So wrote the philosopher John Locke in 1690, in one of the earliest discussions of relativity. Whether the board sits on a table or is carried from here to there makes no difference to how the game is played. As for the chess pieces, so for us. Whether the Earth sits immobile at the center of the cosmos or speeds around the sun, all our activities go on in their customary ways.

Chapter Twenty-Nine

SPUTNIK IN ORBIT, 1687

In a story called "The Red-Headed League," Dr. Watson looks hard at Sherlock Holmes's latest visitor, but nothing strikes him as noteworthy. He turns toward the great detective. Perhaps Holmes has seen more? "Beyond the obvious facts that he has at some time done manual labour, that he takes snuff, that he is a Freemason, that he has been in China, and that he has done a considerable amount of writing lately, I can deduce nothing else," says Holmes.

Galileo and his fellow scientists favored a similar technique. By paying close attention to what others had overlooked, they could find their way to utterly unexpected conclusions. Galileo's analysis of life on shipboard showed, for instance, that a marble that rolled off a table would take precisely the same time to reach the floor whether the ship was moving at a steady speed or standing still. The ship's horizontal motion has no effect on the rock's vertical fall. In Galileo's hands, that seemingly small observation had momentous consequences.

Picture any projectile moving through the air—a baseball soaring toward the outfield, a penny flipped into the air, a dancer leaping across the stage. In all such cases, the moving object's horizontal motion and its vertical motion take place independently and can be examined separately. The horizontal movement is steady and unchanging, in line with Galileo's law of motion. Ball and coin and dancer travel a certain distance horizontally in the first second, the same distance in the next second, and so on, moving at a constant speed from liftoff until touchdown.* At the same time, the projectile's *vertical* progress—its height above the ground—changes according to a different rule. At the moment of launch, the projectile rises quickly but then it rises slower and slower, stops rising altogether, and sits poised for an instant neither rising nor falling, and then plummets earthward faster and faster. The change in speed follows a simple, precise rule, and the upward part of the flight and the downward part are exactly symmetrical.

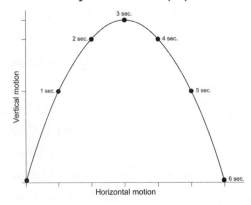

Any object launched into the air—arrow, bullet, cannonball—travels in a curved path like this one. The moving object covers the same horizontal distance during each second of its flight.

Mathematically, it's easy to show that the combination of steady horizontal motion and steadily changing vertical motion

---

* Ballet dancers and basketball players seem to hang in midair, but that is an illusion. The trick for both dancer and athlete is to throw in a few moves midflight. The eye reads the extra motions as taking extra time.

makes for a parabolic path. (A parabola is an arch-shaped curve, but it is not just a generic arch; it is one that satisfies specific technical conditions, just as an ellipse is not a generic oval but one of a specific sort.) Parabolas had been painted against the sky ever since the first caveman threw a rock, but no one before Galileo had ever recognized them, and he was immensely proud of his discovery. "It has been observed that missiles and projectiles describe a curved path of some sort," he wrote. "However no one has pointed out the fact that this path is a parabola. But this and other facts, not few in number or less worth knowing, I have succeeded in proving."

God had once again shown his taste for geometry. The planets in the heavens traveled not in haphazard curves but in perfect ellipses, and objects here on Earth traced exact parabolas.

Concealed within the same observation about the independence of horizontal motion and vertical motion was a further surprise. Galileo might have found it, but he didn't. Isaac Newton did. Imagine someone firing a gun horizontally, and at the same instant someone standing next to the shooter and dropping a bullet from the same height as the gun. When the two bullets reach the ground, they will be far apart. The one shot from the gun will have traveled hundreds of yards; the other will rest in the grass directly below the spot where it was dropped. Which bullet will hit the ground first?

Surprisingly, both reach the ground at exactly the same moment. That's what it means for the bullet's vertical motion— its fall—to be independent of its horizontal motion. For Newton, that was enough to draw a remarkable conclusion.

Suppose it takes one second for a bullet dropped from a

certain height to hit the ground. That means that a bullet shot horizontally from the same height would also hit the ground in one second. A more powerful gun would send the bullet faster and farther, but—if the ground was perfectly flat—that bullet, too, would fall to the ground in one second.

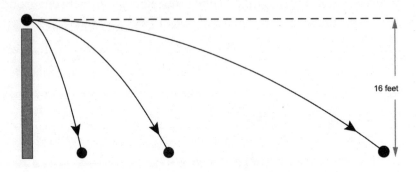

16 feet

Bullets shot horizontally with different force travel different distances before they come to rest, but they all fall at the same rate. Each second a bullet is in the air it falls 16 feet toward the ground.

Newton preferred to imagine a cannon blasting away horizontally. He imagined faster and faster cannonballs, covering greater and greater distances in their one-second journey. But the Earth is round, not flat.

That makes all the difference. Since the Earth isn't flat, it curves away beneath the speeding cannonball. In the meantime, the cannonball is falling toward the ground. Suppose you fired a cannonball from high above the atmosphere, horizontally. With nothing to slow it down, it would continue at the same speed forever, falling all the while. If you launched it at just the right speed, then by the time the cannonball had fallen, say, four feet, the ground itself would have fallen four feet below horizontal.

And then what? The cannonball would continue on its journey forever, always falling but never coming any closer to the ground. Why? Because the cannonball always falls at the same rate, and the ground always curves beneath it at the same rate, so the cannonball falls and falls, and the Earth curves and curves, and the picture never changes. We've launched a satellite.

Newton pictured it all in 1687.

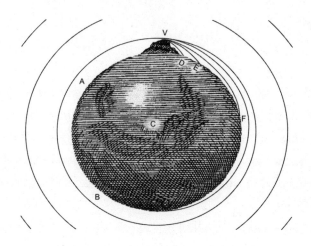

Chapter Thirty

HIDDEN IN PLAIN SIGHT

Kepler had taken the first giant steps toward showing that mathematics governed the heavens. Galileo showed that mathematics reigned here on Earth. Newton's great achievement, to peek ahead for a moment, was to demonstrate that Kepler's discoveries and Galileo's fit seamlessly together, and to explain why.

It was Kepler who spelled out explicitly the credo that all the great seventeenth-century scientists endorsed. When he began studying astronomy he had talked of planets as if they had souls. He soon recanted. The planets surely moved, but their motion had nothing in common with that of galloping horses or leaping porpoises. "My aim is to show that the machine of the universe is not similar to a divine animated being," Kepler declared, "but similar to a clock."

Galileo was the first to grasp, in detail, the workings of the cogs and gears of that cosmic clock. He liked to tell a story, perhaps invented, about how he had made his first great discovery. He had been young and bored, in church, daydreaming. An attendant had lit the candles on a giant chandelier and inadvertently set it swinging. Rather than listen to the service, Galileo watched the chandelier. It swung widely at first and then gradually in smaller and smaller arcs. Using his pulse beat to measure the time (in his day no one had yet built a clock with a second hand), Galileo discovered what has ever since been known as the law of pendulums—a pendulum takes the same time to swing through a small arc as through a large one.

Perhaps it was because Galileo had been raised in a musical household—his father was a renowned composer and musician—that counting time came naturally to him.* Eventually his counting would lead to one of history's profound discoveries. What Galileo did, and what no one before him had ever done, was find a new way to think about time. It was an accomplishment akin to a fish's finding a new way to think about water. "Galileo spent twenty years wrestling with the problem before he got free of man's natural biological instinct for time as that in which he lives and grows old," wrote the historian Charles C. Gillispie. "Time eluded science until Galileo."

Galileo's solution was so successful and so radical that everyone today—even those without the slightest knowledge of physics—takes his insight for granted. The breakthrough was to identify time—not distance or temperature or color or any of a thousand other possibilities—as the essential variable that governs the world. For years Galileo had tried to find a relationship between the speed of a falling object and the distance it had fallen. All his efforts failed. Finally he turned away from distance and focused on time. Suddenly everything fell into place. Galileo had found a way to pin numbers to the world.

The crucial experiments might have occurred only to a musician. Once again they involved rolling a ball down a ramp. The

---

* "Music," Leibniz wrote, "is the pleasure the human soul experiences from counting without being aware that it is counting."

setup was bare-bones: a wooden ramp with a thin groove down the middle, a bronze ball to roll down the groove, and a series of movable catgut strings. The strings lay on the surface of the ramp, at a right angle to the groove, like frets on the neck of a guitar. When the ball crossed a string, it made an audible click but its speed continued almost unchanged.

Galileo may actually have dropped rocks from the Leaning Tower of Pisa, as legend has it, but if he did they fell too quickly to study. So he picked up a ball, released it at the top of the ramp, and cocked his ears.

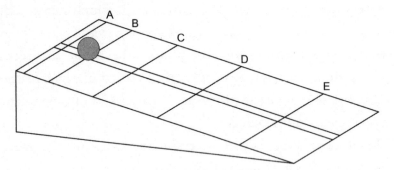

Now the strings came into play. Galileo could hear the ball cross each string in turn, and he painstakingly rolled the ball again and again, each time trying to position the strings so that the travel time between each pair of strings was the same. He needed to arrange the strings, in other words, so that the time it took the ball to move from the top of the ramp to string A was the same as the time it took to move from string A to string B, which was the same as the time from B to C, and C to D, and so on. (He measured time intervals by weighing the water that leaked through a hole in the bottom of a jug. Twice as much water meant twice as much time.) It was finicky, tedious work.

Finally satisfied, Galileo measured the distance between strings. That yielded this little table.

	Time (in seconds)	Distance (in inches)
Start to A	1	1
A to B	1	3
B to C	1	5
C to D	1	7
D to E	1	9

The pattern in the right-hand column was easy to spot, but Galileo looked at the numbers again and recast the same data into a new table. Instead of looking at the ball as it traveled from one string to the next, he focused on the total distance the ball had traveled from the starting line. (All he had to do was add up the distances in the right-hand column.) This time he saw something more tantalizing.

	Time (in seconds)	Distance (in inches)
Start to A	1	1
Start to B	2	4
Start to C	3	9
Start to D	4	16
Start to E	5	25

Each number in the right column of this new table represented the distance the ball had traveled in a certain amount of time—in one second, in two seconds, in three seconds, and so on. That distance, Galileo saw, could be expressed as a function of time. In t seconds, a ball rolling down a ramp at gravity's command traveled precisely t^2 inches.* In 1 second, a ball rolled 1^2 inches, in 2 seconds 2^2 inches, in 5 seconds 5^2 inches, and so on.

What was just as surprising was what the law *didn't* say—it didn't say anything about how much the ball weighed. Roll a cannonball and a BB down a ramp in side-by-side grooves, and they would travel alongside one another all the way and reach the bottom at precisely the same moment. For a given ramp, the same tidy law always held—the distance the ball traveled was proportional to time squared. All that counted was the height above the ground of the point where the ball was released.

Repeat the experiment on a steeper ramp, and the cannonball and the BB would both travel faster, but they would still travel side by side every inch of the way. That was enough. Galileo made a daring leap: what held for a steep ramp and for an even steeper ramp would also hold for the steepest "ramp" of all, a free fall through the air. All objects, regardless of their weight, fall at exactly the same rate.

---

* To be more accurate, in t seconds a ball falls a distance *proportional* to t^2 inches rather than precisely equal to t^2 inches. (It falls, for instance, $3 \times t^2$ inches or $10 \times t^2$ inches or some other multiple, depending on the steepness of the ramp.) Everything I've said here carries over to the more general case, but the numbers would be off-putting. For purposes of illustration, I chose the ramp that showed the pattern most clearly.

Chapter Thirty-One

TWO ROCKS AND A ROPE

Tradition has it that Galileo discovered how objects fall by dropping weights from the top of the Leaning Tower of Pisa. Unlike most legends—Archimedes and his bathtub, Columbus and the flat Earth, George Washington and the cherry tree—historians believe this one might possibly be true. A tower drop would have disproved Aristotle's claim that heavy objects fall faster than light ones. But it took the ramp experiments, which Galileo indisputably carried out, to yield the quantitative law about distance and time.

Whether he really climbed a tower or not, Galileo did propose a thought experiment to test Aristotle's claim. Imagine for a moment, said Galileo, that it was true that the heavier the object, the faster its fall. What would happen, he asked, if you tied a small rock and a big rock together, with some slack in the rope that joined them? On the one hand, the tied-together rocks would fall *slower* than the big rock alone, because the small rock would lag behind the big one and bog it down, just as a toddler tied to a sprinter would slow him down. (That was where the slack in the rope came into play.) On the other hand, the tied-together rocks would fall *faster* than the big rock alone, because they constituted a new, heavier "object."

Which meant, Galileo concluded triumphantly, that Aristotle's assumption led to an absurd conclusion and had to be abandoned. Regardless of what Aristotle had decreed, logic forced us to conclude that all objects fall at the same rate, regardless of their weight. This is a story with a curious twist. Galileo, the great pioneer of experimental science, may never have bothered to perform his most famous experiment. No one is sure. What we know with certainty is that, like the Aristotelians he scorned, Galileo sat in a chair and deduced the workings of the world with no tool but the power of logic.

Since Galileo's day, countless tests have confirmed his Leaning Tower principle (including some at the Leaning Tower itself). In ordinary circumstances, air resistance complicates the picture— feathers flutter to the ground and arrive long after cannonballs. Not until the invention of the air pump, which came after Galileo's death, could you drop objects in a vacuum. A century after Galileo, the demonstration retained its power to surprise. King George III demanded that his instrument makers arrange a test for him, featuring a feather and a one-guinea coin falling in a vacuum. "In performing the experiment," one observer wrote, "the young optician provided the feather, the King supplied the guinea and at the conclusion the King complimented the young man on his skill as an experimenter but frugally returned the guinea to his waistcoat pocket."

Today we've all seen the experiment put to the test, at every Olympic games. When television shows a diver leaping from the ten-meter board, thirty feet above the pool, how does the camera stick with her as she plummets toward the water? Galileo could have solved the riddle—just as a small stone falls at exactly the same rate as a heavy one, a camera falls at exactly the same rate as a diver. The trick is to set up a camera near the diver, at exactly the same height above the water. Attach the camera to a vertical pole and release the camera at the instant the diver starts her fall poolward. Gravity will do the rest.

Galileo exulted in his discovery that "distance is proportional to time squared." The point was not merely that nature could be described in numbers but that a single, simple law—in operation since the dawn of time but unnoticed until this moment (just as the Pythagorean theorem had been true but unknown before *its* discovery)—applied to the infinite variety of falling objects in the world. A geranium knocked off a windowsill, a painter tumbling off his ladder, a bird shot by a hunter, all fell according to the same mathematical law.

The difference between Galileo's world and Aristotle's leaps out, as we have seen. Galileo had stripped away the details that fascinated Aristotle—the color of the bird's plumage, the motives behind the painter's absentmindedness—and replaced the sensuous, everyday world with an abstract, geometric one in which both a bird and a painter were simply moving dots tracing a trajectory against the sky. Ever since, we have been torn between celebrating the bounty that science and technology provide and lamenting the cost of those innovations.

Chapter Thirty-Two

A FLY ON THE WALL

The mathematical patterns that Kepler had found in the heavens looked different from those Galileo had found on Earth. Perhaps that was to be expected. What did falling rocks have to do with endlessly circling planets, which plainly were not falling at all?

Isaac Newton's answer to that question would make use of mathematical tools that Kepler and Galileo did not know. Both astronomers were geniuses, but everything they found might conceivably have been discovered in Greece two thousand years before. To go further would require a breakthrough the Greeks never made.

The insight that eluded Euclid and Archimedes (and Kepler and Galileo as well) supposedly came to René Descartes when he was lying in bed one morning in 1636, idly watching a fly crawl along the wall. ("I sleep ten hours every night," he once boasted, "and no care ever shortens my slumber.") The story—so claimed one of Descartes' early biographers—was that Descartes realized that the path the fly traced as it moved could be precisely described in numbers. When the fly first caught Descartes' eye, for instance, it was 10 inches above the floor and 8 inches from the left-hand edge of the wall. A moment later it was 11 inches

above the floor and 9 inches from the left edge. All you needed
were two lines at right angles—the horizontal line where the
wall met the floor, say, and the vertical line from floor to ceiling where two walls met. Then at any moment the fly's position
could be pinpointed—this many inches from the horizontal line,
that many from the vertical.

Pinpointing a location was an old idea, as old as latitude and
longitude. The new twist was to move beyond a static description
of the present moment—the fly is 11 inches from here, 9 inches
from there; Athens is at 38°N, 23°E—and to picture a *moving*
point and the path it drew as it moved. Take a circle. It can be
thought of in a static way, as a particular collection of points—
all those points sitting precisely one inch from a given point, for
instance. Descartes pictured circles, and other curves, in a more
dynamic way. Think of an angry German Shepherd tethered to
a stake and straining to reach the boys teasing him, just beyond
his reach. The dog traces a circle—or, more accurately, an arc
that forms part of a circle—as he moves back and forth at the
end of his taut leash. A six-year-old on a swing, pumping with
all his might, traces out part of a circle as the swing arcs down
toward the ground and then up again.

From the notion of a curve as a path in time, it was but a step
to the graphs that we see every day. The key insight was that the
two axes did not necessarily have to show latitude and longitude;
they could represent *any* two related quantities. If the horizontal
axis depicted "time," for instance, then a huge variety of numerical changes suddenly took on pictorial form.

The most ordinary graph—changes in housing prices over
the last decade, rainfall this year, unemployment rates for the
past six months—is an homage to Descartes. A table of numbers might contain the identical information, but a table muffles

the patterns and trends that leap from a graph. We have grown
so accustomed to graphs that show how something changes as
time passes that we forget what a breakthrough they represent.
(Countless expressions take this familiarity for granted: "off the
charts," "steep learning curve," "a drop in the Dow.") Any run-
of-the-mill illustration in a textbook—a graph of a cannonball's
position, moment by moment, as it flies through the air, for
example—is a sophisticated abstraction. It amounts to a series
of stop-action photos. No such photos would exist for centuries
after Descartes' death. Only familiarity has dulled the surprise.*

Even in its humblest form (in other words, even aside from
thinking of a curve as the trajectory of a moving point), Descartes' discovery provided endless riches. With his horizontal
and vertical axes in place, he could easily construct a grid—he
could, in effect, tape a piece of graph paper to any spot he
wanted. That assigned every point in the world a particular ad-
dress: x inches from this axis, y inches from that one. Then, for
the first time, Descartes could approach geometry in a new way.
Rather than think of a circle, say, as a picture, he could treat it as
an equation.

A circle consisted of all the points whose x's and y's combined
in a particular way. A straight line was a different equation, a
different combination of x's and y's, and so was every other curve.

---

* One prominent historian calls it "incomprehensible" that Greek mathematicians never conceived of graphs. But neither did their intellectual
descendants for well over a thousand years. Even an enormous hint went
unnoticed. Monks in the Middle Ages invented musical notation, which
meant they no longer had to commit countless chants to memory. "The
musical staff was Europe's first graph," noted the historian Alfred Crosby,
but several more *centuries* would pass before scientists saw that they, too,
could use graphs to depict changes in time.

A curve was an equation; an equation was a curve. This was a huge advance, in the judgment of John Stuart Mill "the greatest single step ever made in the progress of the exact sciences." Now, suddenly, all the tools of algebra—all the well-developed arsenal of techniques for manipulating equations—could be enlisted to solve problems in geometry.

But it was not simply that algebra could be brought to bear on geometry. That would have been a huge practical breakthrough, but Descartes' insight was a conceptual revolution as well. Algebra and geometry had always been seen as independent subjects. The distinction wasn't subtle. The two fields dealt with different topics, and they looked different. Algebra was a forest of symbols, geometry a collection of pictures. Now Descartes had come along and showed that algebra and geometry were two languages that described a shared reality. This was completely unexpected and hugely powerful, as if today someone suddenly showed that every musical score could be converted into a scene from a movie and every movie scene could be translated into a musical score.

Chapter Thirty-Three

"EUCLID ALONE HAS LOOKED ON BEAUTY BARE"

Descartes unveiled his new graphs in 1637, in an appendix to a work called *Discourse on Method*. The book is a milestone in the history of philosophy, the source of one of the best known of all philosophical maxims. In the *Discourse* Descartes set out his determination to reject all beliefs that could possibly be incorrect and to build a philosophy founded on indisputable truths. The world and everything in it might be an illusion, Descartes argued, but even if the world was but a dream it was *his* dream, and so he himself could not be merely an illusion. "I think, therefore I am."

In the same work he added three short afterwords, each meant to demonstrate the power of his approach to philosophy. In an essay called "Geometry," Descartes talked about curves and moving points; he explained that a curve can be depicted in a picture or captured in an equation and showed how to translate between the two; he discussed graphs and the use of what are known today as Cartesian coordinates. He understood the value of what he had done. "I do not enjoy speaking in praise of myself," he wrote in a letter to a friend, but he forced himself. His new, graph-based approach to geometry, he went on, represented a leap "as far beyond the treatment in the ordinary geometry as the rhetoric of Cicero is beyond the ABC of children."

It did. The wonder is that something so useful and so obvious—in hindsight—should have eluded the world's greatest thinkers for thousands of years. But this is an age-old story. In the making of the modern world, the same pattern has recurred time and again: some genius conceives an abstract idea that no one before had ever grasped, and in time it finds its way so deeply into our lives that we forget that it had to be invented in the first place.

Abstraction is always the great hurdle. Alfred North Whitehead argued that it was "a notable advance in the history of thought" when someone hit on the insight that two rocks and two days and two sticks all shared the abstract property of "twoness." For countless generations no one had seen it.

The same holds for nearly every conceptual breakthrough. The idea that "zero" is a number, for instance, proved even more elusive than the notion of "two" or "seven." Whitehead again: "The point about zero is that we do not need to use it in the operations of daily life. No one goes out to buy zero fish. It is in a way the most civilized of all the [numbers], and its use is only forced on us by the needs of cultivated modes of thought." With zero in hand, we suddenly have a tool kit that lets us start building the conceptual world. Zero opens the way to place notation—we can distinguish 23 from 203 from 20,003—and to arithmetic and algebra and countless other spinoffs.

Negative numbers once posed similar mysteries. Today the concept of a $5 bill is easy to understand, and so is a $5 IOU.

A temperature of 10 degrees is straightforward, and so is 10 degrees below zero. But in the history of the human race, for the greatest intellects over the course of millennia, the notion of negative numbers seemed as baffling as the idea of time travel does to us. (Descartes wrestled to make sense of how something could be "less than nothing.") Numbers named amounts—1 goat, 5 fingers, 10 pebbles. What could negative 10 pebbles mean?

(Lest we grow too smug we should remember the dismay of today's students when they meet "imaginary numbers." The name itself [coined by Descartes, in the same essay in which he explained his new graphs] conveys the unease that surrounded the concept from the start. Small wonder. Students still learn, by rote, that "positive times positive is positive, and negative times negative is positive." Thus, $-2 \times -2 = 4$, and so is 2×2. Then they learn a new definition—an imaginary number is one that, when multiplied by itself, is *negative*! It took centuries and the labors of some of the greatest minds in mathematics to sort it out.)

The ability to conceive strange, unintuitive concepts like "twoness" and "zero fish" and "negative 10 pebbles" lies at the heart of mathematics. Above all else, mathematics is the art of

Reality versus abstraction. Photo of cow, left. Painting of cow by Dutch artist Theo van Doesburg, right, © *The Museum of Modern Art/licensed by SCALA/Art Resources, NY.*

abstraction. It is one thing to see two apples on the ground next to three apples. It is something else to grasp the universal rule that $2 + 3 = 5$.

In the history of science, abstraction was crucial. It was abstraction that made it possible to look past the chaos all around us to the order behind it. The surprise in physics, for instance, was that nearly everything was beside the point. Less detail meant more insight. A rock fell in precisely the same way whether the person who dropped it was a beauty in silk or an urchin in rags. Nor did it matter if the rock was a diamond or a chunk of brick, or if it fell yesterday or a hundred years ago, or in Rome or in London.

The skill that physics demanded was the ability to look past particulars to universals. Just as someone working on a geometry problem would not care whether a triangle was drawn in pencil or ink, so a scientist seeking to describe the world would dismiss countless details as true but irrelevant. Much of a modern physicist's early training consists in learning to transform colorful questions about such things as elephants tumbling down mountainsides into abstract diagrams showing arrows and angles and masses.

The move from elephants to ten-thousand-pound masses echoes the transformation from Aristotle's worldview to Galileo's. The battle between the two approaches was as sweeping as a contest can be, far more than a debate over whether the sun circled the Earth or vice versa, big as that issue was. The broader questions had to do with how to study the physical world. For Aristotle and his followers, the point of science was to engage with the real world in all its complexity. To talk of weights plummeting through vacuums or perfect spheres rolling forever across infinite planes was to mistake idealized diagrams for reality. But the map was not the territory. Explorers needed to grapple with the world as it is, not with a dessicated and lifeless counterpart.

In Galileo's view, this was exactly backward. The way to understand the world was not to focus on its every quirk and blemish but to look beyond those distractions to the deeper truths they obscured. When Galileo talked about whether heavy objects fall faster than light ones, for instance, he imagined ideal circumstances—objects falling in a vacuum rather than through the air—in order to avoid the complications posed by air resistance. But Aristotle insisted that no such thing as a vacuum could exist in nature (it was impossible, because objects fall faster in a thin medium, like water, than they do in a thick one, like syrup. If there were vacuums, then objects would fall infinitely fast, which is to say they would be in two places at once).* Even if a vacuum could somehow be contrived, why would anyone think that the behavior of objects in those peculiar conditions bore any relation to ordinary life? To speculate about what might happen in unreal circumstances was an exercise in absurdity, like debating whether ghosts can get sunburns.

Galileo vehemently disagreed. Abstraction was not a distortion but a means of seeing truth unadorned. "Only by imagining an impossible situation can a clear and simple law of fall be

* The question of whether vacuums could exist spurred long, angry debates. The invention of the air pump did not settle the debate, in the view of Leibniz and some others, because even if a jar no longer contained *air* it might still contain some more ethereal fluid. Leibniz and Descartes both maintained that the very notion of a vacuum was nonsensical—how could there be a place containing nothing at all, when the meaning of the word *place* is "the location where something happens to be"? Newton and Pascal insisted just as vehemently that vacuums were real. Descartes contended, cattily, that the only vacuum was in Pascal's head.

formulated," in the words of the late historian A. Rupert Hall, "and only by possessing that law is it possible to comprehend the complex things that actually happen."

By way of explaining what the abstract, idealized world of mathematics has to do with the real world, Galileo made an analogy to a shopkeeper measuring and weighing his goods. "Just as the accountant who wants his calculations to deal with sugar, silk, and wool must discount the boxes, bales and other packings, so the mathematical scientist . . . must deduct the material hindrances" that might entangle him.

The importance of abstraction was a crucial theme, and Galileo came back to it often. At one point he exchanged his shopkeeper image for a more poetic one. With abstraction's aid, he wrote, "facts which at first sight seem improbable will . . . drop the cloak which has hidden them and stand forth in naked and simple beauty."

Galileo won his argument, and science has never turned back. Mathematics remains the language of science because, ever since Galileo, we have taken for granted that abstraction is the pathway to truth.

HERE BE MONSTERS!

Science was now poised to confront one of its great taboos. The study of objects in motion had tempted and intimidated thinkers since ancient times. With his work on ramps and his discovery of the law of falling objects, Galileo had mounted the first successful assault. With his insights into graphs and the curves traced by moving points, Descartes had devised the tools that would make an all-out attack possible. Only one giant obstacle still blocked the way.

How did it happen that the Greeks, whose intellectual daring has never been surpassed, shied away from applying mathematics to objects moving through space? In part because, as we have seen, they deemed impermanence an unworthy subject for mathematics, which investigated eternal truths. But they were skittish, too. That uneasiness was largely due to one man, named Zeno, who lived in a middle-of-nowhere Greek colony in southern Italy sometime around 450 B.C. Zeno figured in one of Plato's *Dialogues* (Plato called him "tall and fair to look upon"), but almost all the facts about his life have been lost to us. So has almost every scrap of his writing. The few snippets that have survived have tied philosophers in knots from his day to ours.

Zeno's arguments sound silly at first, almost childish, for he was one of those philosophers who spoke not in polysyllables and abstractions but in stories. Only four of those tales have survived. Each is a tiny, paradoxical fable, a Borges parable unstuck in time by two millennia.

One story starts with a man standing in a room. His goal is to walk to the far side. Could anything be simpler? But before he can cross the room, Zeno points out, the man must first reach the halfway point. That will take a small but definite amount of time. And then he must cross half the distance that remains. Which will take a certain amount of time. Then half the still-remaining distance, and so on, forever. "Forever" is the key. A trip across the room, then, must pass through an infinite number of stages, each of which takes some definite, more-than-zero amount of time. And that could only mean, Zeno concluded gleefully, that a trip across the room would necessarily take an infinite amount of time.

Zeno certainly didn't believe that a man in a room was doomed to die before he could reach a doorway on the other side. His challenge to his fellow philosophers was not to cross a room but to find a mistake in his reasoning. On the one hand, everyone knew how to walk from here to there. On the other hand, that seemed impossible. What was going on?

For two thousand years, no one came up with a satisfactory answer. Philosophers debated endlessly, for instance, whether it even made sense to talk of dividing time into ever tinier bits—is time continuous, like a ribbon, or is it more like a series of beads on a string? Can time be divided up forever or does it come in irreducible units, like atoms?

The Greeks quit in frustration early on. They did note that each of Zeno's tales began as a commonplace story having to do with motion and ended up circling around the strange notion of infinity. The danger zone seemed clear enough. Motion meant infinity, and infinity meant paradox. Having failed to find Zeno's error, Greek mathematicians opted to do the prudent thing. They put up emergency cones and yellow police tape and made a point of staying well clear of anything that involved the analysis of moving objects. "By instilling into the minds of the Greek geometers a horror of the infinite," the mathematician Tobias Dantzig observed, "Zeno's arguments had the effect of a partial paralysis of their creative imagination. The infinite was taboo, it had to be kept out, at any cost."

That banishment lasted twenty centuries.

Occasionally, during the long hiatus, a particularly bold thinker tiptoed to the brink of infinity, glanced down, and then hurried away. Albert of Saxony, a logician who lived in the 1300s, was one of the most insightful of this small band. To demonstrate just how strange a concept infinity is, Albert proposed a thought experiment. Imagine an infinitely long wooden beam, one inch high and one inch deep. Now take a saw and cut the beam into identical one-inch cubes. Since the beam is infinitely long, you can cut an infinite number of cubes.

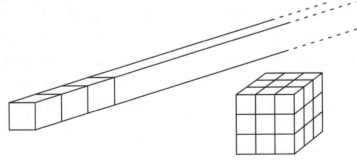

An infinitely long beam can be cut into blocks and then reassembled into bigger and bigger cubes.

What Albert did next was as surprising as a magic trick. The original beam, with a cross section of only one square inch, certainly did not take up much room. It went on forever, but you could easily hop over it. But if you took cubes from that beam and arranged them cleverly, Albert showed, you could fill the entire universe to the brim. The scheme was simple enough. All you had to do was build a small cube and then, over and over again, build a bigger cube around it.

First, you set a single cube on the ground. Then you made a $3 \times 3 \times 3$ cube with the original cube at its center. That cube in turn became the center of a $5 \times 5 \times 5$ cube, and so on. In time, the skinny beam that you began with would yield a series of colossal cubes that outgrew the room, the neighborhood, the solar system!

Once again, the moral was plain. To explore infinity was to tumble into paradox. Like the Greeks fifteen centuries before, medieval mathematicians edged away from the abyss.

Three hundred years later, Galileo ventured back toward forbidden territory. He began so innocuously that it seemed impossible he could fall into danger. Consider, Galileo said, one of the humblest of all intellectual activities: matching, a skill even more primitive than counting. How can we tell if two collections are the same size? By taking one item from the first collection and matching it with one from the other collection. Then we set those two aside and start over. How do we know that there are five vowels? Because we can match them up with our five fingers—the letter *a* with the thumb, say, and *e* with the index finger, *i* with the middle finger, *o* with the ring finger, and *u* with the pinky. Each vowel pairs off with a finger; each finger pairs off with a vowel; no member of either group is left over or left out.

Let us pause to make one more observation, which also seems utterly obvious. If we think of a group—everyone who lives in Italy, say—and then we think of a smaller group contained within it—everyone who lives in Rome—then it seems beyond question that the original group is bigger than the subgroup.

In a moment, we will see why these points are worth belaboring. Suppose, said Galileo, we take not simply a big group, like the citizens of Italy, but an infinite group, like the counting numbers. Galileo wrote them in a line like this:

$$1 \; 2 \; 3 \; 4 \ldots$$

Next, said Galileo, suppose we think of a smaller group contained within the large one. Take, for instance, the numbers 1^2, 2^2, 3^2, 4^2, and so on. (In other words, the numbers 1, 4, 9, 16 . . .). Galileo wrote them in a line of their own:

$$1^2 \; 2^2 \; 3^2 \; 4^2 \ldots$$

Then he sprung his trap. Since the list 1, 4, 9, 16 . . . plainly leaves out a great many numbers, it is beyond question smaller than the collection of *all* numbers. But Galileo arranged the two lines of numbers one below the other and paired them up tidily.

Each number in the top line had exactly one partner in the

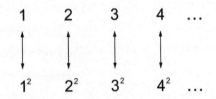

bottom line, and vice versa. Every number had a partner; no number had more than one partner; and no number in either line was left out. (The reason Galileo chose the numbers 1^2, 2^2, 3^2 . . . in the first place was that they could so readily be paired up with

1, 2, 3 . . .) That left only one conclusion. Since the two collections matched exactly, they were the same size. Galileo had found a working definition of infinity—a collection is infinitely big if part of it is the same size as the whole thing!

Few thinkers in history have been as bold as Galileo, but even for him this was too much. For the sake of an idea, he would one day challenge the Inquisition. But faced with the paradoxes of infinity, he blinked and hurried away.

Galileo, brilliant in so many domains, had pinpointed infinity's strangest property. In a sense, big numbers are all alike. One million is bigger than one thousand, but you can get from one to the other. All it takes is patience. Add one. Add one more. Add another. Eventually you get there. But infinity sits on the far side of a chasm that you can never bridge. When it comes to infinity, it's not only that another (and another and another) doesn't bring you *to* the goal; worse than that, it doesn't bring you *any nearer* to the goal.

That idea, so remote from anything in the everyday world, continues to baffle even the deepest thinkers. In *Portrait of the Artist as a Young Man*, James Joyce took a stab at conveying the notion of infinity. The damned suffer eternally in hell. "For ever! For all eternity!" Joyce wrote. "Not for a year or for an age but for ever. Try to imagine the awful meaning of this. You have often seen the sand on the seashore. How fine are its tiny grains! And how many of those tiny little grains go to make up the small handful which a child grasps in its play. Now imagine a mountain of that sand, a million miles high, reaching from the earth to the farthest heavens, and a million miles broad, extending to remotest space, and a million miles in thickness. . . ."

On and on Joyce went, the limitlessly talented writer multiplying grains of sand by drops of water in the ocean by stars in the sky. And *still* he came up short, still he failed to narrow the gap between the finite and the infinite. Because the essential point is that infinity is not just a big number but something much, much more bizarre than that.

Isaac Newton was one of the greatest of all geniuses and one of the strangest men who ever lived. "The most fearful, cautious, and suspicious Temper that I ever knew," in one contemporary's words, he was born sickly and premature but lived to 84 (and died a virgin). Born on Christmas Day, Newton believed with all his heart that he had been selected by God to decode His secrets. This portrait shows him at the peak of his powers, at age 46, just after he had unveiled the theory of gravitation.

Charles II, whose father had been beheaded in 1649, ruled England from 1660 to 1685. Witty and restless, the "Merrie Monarch" presided over a self-indulgent court in which everyone, from the king on down, was "engaged in an endless game of sexual musical chairs." Science, too, fascinated Charles. He founded the Royal Society and poses here with a telescope and other scientific instruments.

Robert Boyle, an aristocrat whose father was one of Britain's richest men, was the most-respected member of the Royal Society in its early days. Boyle was a brilliant scientist who also believed, among other things, that the best cure for cataracts was to blow powdered, dried, human excrement into the patient's eyes.

Gottfried Leibniz was Newton's great rival and himself a man of astonishing genius. As energetic as a puppy, Leibniz was the temperamental opposite of the austere Newton. Leibniz was a philosopher, a mathematician, an inventor, and a man so pleased with himself that his favorite wedding gift to new brides was a collection of his own maxims.

The 1600s saw the birth of science and the modern age, but old beliefs and fears still held prominent places in men's hearts. Comets were dreaded, as they had been for ages. This scene from the Bayeux tapestry (stitched in the 11th century) shows men cowering in fear as Halley's Comet appears overhead, in 1066.

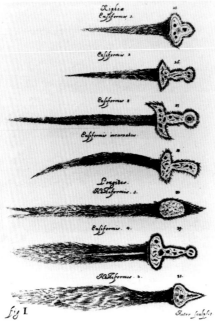

When our forebears looked at comets, they saw not merely glowing lights but fiery, death-wielding swords, as in these illustrations from 1668. Comets had appeared overhead in 1664 and 1665, and England braced for "a MORTALITY which will bring MANY to their Graves."

In 1665, plague struck England. No one knew the cause, and no one knew a cure. Somehow the killer leaped from victim to victim. Many were healthy one day and dead that night. Plague doctors, shown here, could offer little but a kind word. Their costume was meant to protect the wearer; the beak contained herbs and spices, to counter the smell of the dead and dying.

Bills of Mortality recorded the weekly death toll. The first deaths in London came at a rate of one or two per month. At its peak, in September 1665, plague killed more than six thousand Londoners in a single week. The stricken city was nearly silent except for the tolling of church bells.

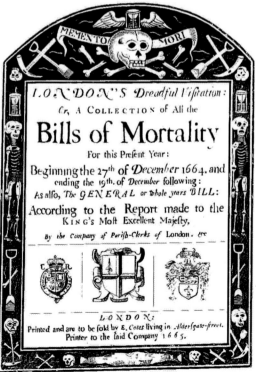

MEMENTO MORI

LONDON'S Dreadful Visitation:

Or, A COLLECTION of All the

Bills of Mortality

For this Present Year:

Beginning the 27th of *December* 1664. and ending the 19th. of *December* following:

As also, The GENERAL or whole years BILL:

According to the Report made to the KING's Most Excellent Majesty,

By the Company of Parish-Clerks of London. &c

LONDON:

Printed and are to be sold by E. *Cotes* living in *Aldersgate-street*, Printer to the said Company 1 6 6 5.

Samuel Pepys lived through the turmoil of London in the late 1600s and took notes in his diary, in a secret code. An administrator in the British navy, he reported on office politics, dalliances with mistresses, and fights with his wife. His accounts of the plague and the Great Fire of London are the most detailed descriptions we have.

As soon as the plague released England from its chokehold, a second calamity swooped down. In the year 1666—ominous because of the fearsome number 666—London caught fire. For four days the city burned. Iron bars in prison cells melted. One hundred thousand people were left homeless. Plainly an outraged God had lost patience with his creation.

In this harsh era, punishments were carried out in public, the better to warn and entertain. Drawing and quartering was the most gruesome. A man was hanged by the neck but not killed. In England he was then disemboweled (while alive) and cut in quarters. The head and body parts were nailed up around the city. In France, as in this picture, horses performed the quartering.

Spectators out for a day's entertainment might attend a puppet show or a hanging or, perhaps, a bear-baiting. Dogs attacked a chained bear, who flailed at his attackers. Bull-baiting was popular, too. The sport gave rise to the English bulldog, whose flat face made it possible for him to keep breathing without releasing his hold on the bull.

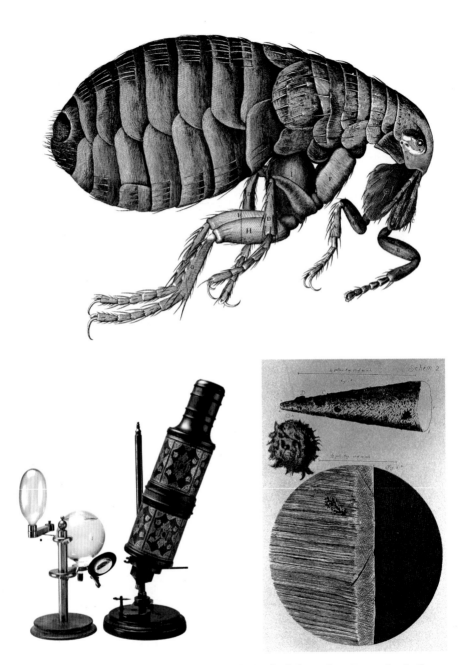

The microscope astonished all those who looked through it. Even a lowly flea, like this one drawn by Robert Hooke, revealed God's perfect artistry. But when Hooke turned his microscope on man-made objects, they looked shoddy in comparison. The point of a needle (at right, top) was rough and pitted, as was a razor's edge (at right, bottom). A printed dot on a page (at right, middle) was "quite irregular, like a great Splatch of London dirt."

Anything was still possible in science's early days. Travelers told wondrous tales of such oddities as Indians in the New World who had no heads but sported eyes on their torsos. Isaac Newton, a devout believer in alchemy, drew the image at the bottom of the page. It refers to a magical substance called the philosopher's stone, which could change ordinary objects to gold and convey immortality. Newton included instructions for coloring the diagram.

Tycho Brahe was the most eminent astronomer of the generation before Kepler and Galileo. Rich and eccentric, Tycho presided over a magnificent observatory on a private island. He had lost part of his nose in a duel and ever after sported a replacement made of gold and silver.

Johannes Kepler was an astronomer, an astrologer, and a mathematical genius. Kepler spent decades studying the astronomical data that Tycho had gathered, in search of patterns that he fervently believed God had hidden like secret messages in a text.

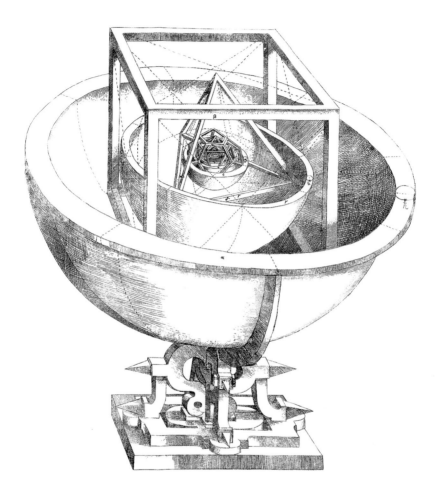

Kepler's proudest achievement was devising an elaborate model that explained how God had laid out the solar system. God was a mathematician, Kepler believed, and He had arranged the planets' orbits in keeping with this intricate geometric model. The sun sat at the center, inside a nested cage of cubes, pyramids, octahedrons, and so on. Beautiful as the model was, it eventually became clear that it had nothing to do with reality.

Galileo was brilliant and pugnacious. A mathematician and astronomer, he amazed Italy's rulers by letting them look out to sea through a telescope, a brand-new invention that he had considerably improved. When Galileo turned his telescope to the skies, he revolutionized our picture of the heavens. Among other discoveries, he found that the moon was not a perfect sphere, as everyone had believed, but was rough and pockmarked. Galileo painted these watercolors himself.

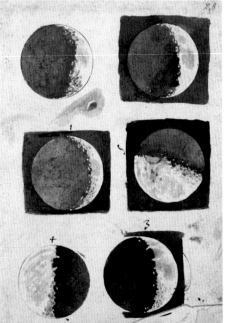

The crucial belief of Isaac Newton and his fellow scientists was that God had designed the world on mathematical lines. All nature followed precise laws. The belief derived from the Greeks, who had been amazed to find that music and mathematics were deeply intertwined. Here they explore the relation between the weight of a bell, or the volume of a glass, and its pitch. Pythagoras, shown below, is credited with being the first to find this connection, "one of the truly momentous discoveries in the history of mankind."

Newton's theory of gravity propelled him to instant fame. He liked to tell the story, depicted in this Japanese print, that the crucial insight came from watching an apple fall. The story is quite likely a myth. Before anyone knew of his mathematical genius, Newton had dazzled the Royal Society with this compact yet powerful telescope.

Edmond Halley (known today for Halley's Comet) was a brilliant astronomer and, just as surprisingly, a man so congenial that he could get along with Isaac Newton. Halley took on the task of coaxing the reluctant, secretive Newton into publishing his masterpiece, *Principia Mathematica*. The 500-page book, in Latin and dense with mathematics, might never have appeared without Halley's labors.

PHILOSOPHIÆ
NATURALIS
PRINCIPIA
MATHEMATICA.

Autore *J S. NEWTON*, *Trin. Coll. Cantab. Soc.* Matheseos
Professore *Lucasiano*, & Societatis Regalis Sodali.

IMPRIMATUR·
S. PEPYS, *Reg. Soc.* PRÆSES.
Julii 5. 1686.

LONDINI,

Jussu *Societatis Regiæ* ac Typis *Josephi Streater*. Prostant Vena-
les apud *Sam. Smith* ad insignia Principis *Walliæ* in Cœmiterio
D. *Pauli*, aliosq; nonnullos Bibliopolas. *Anno* MDCLXXXVII.

Newton's tomb, in Westminster Abbey. From the moment he unveiled the theory of gravity Newton was hailed as almost superhuman. Voltaire observed Newton's funeral and was stunned to see dukes and earls carrying the casket. "I have seen a professor of mathematics, simply because he was great in his vocation, buried like a king who had been good to his subjects."

Chapter Thirty-Five

BARRICADED AGAINST THE BEAST

The closer mathematicians looked at infinity, the stranger it seemed. Take one of the simplest drawings imaginable, a straight line one inch long. That line is made up of points, and there are infinitely many of them. Now draw a line two inches long. The longer line must have twice as many points as the shorter one (what else could make it twice as long?). But a matching technique, much like Galileo had used with numbers, shows that there are precisely the same number of points on both lines.

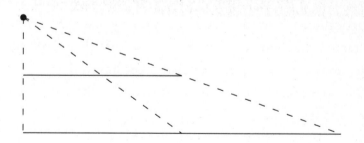

The proof is pictorial. Make a dot as in the drawing and draw a straight line from it through the two lines. Any such line pairs up a point on the short line with a point on the longer line. Like a perfectly orderly dance, everyone has a partner. No point on either line is left out, and no point has to share a partner with anyone else. How can that be?

Worse was to come. Exactly the same argument shows that a line ten inches long is made of precisely as many points as a line one inch long. So is a line ten miles long, or ten thousand. Could anything send a clearer message that infinity was a topic best left to philosophers and mathematicians, and completely unsuited to hardheaded scientists?

Infinity is built into mathematics from the beginning, because numbers go on forever. If someone made a claim about all the human beings on Earth—no person alive today is nine feet tall—in principle you could test it by gathering everyone into a line and working your way along from first person to last. But no such test can work for numbers, because the line never ends. For every number there is another, bigger number (and also another half as big).

But it was by no means clear that infinity had anything to do with the real world. That was fine. Seventeenth-century scientists, like all their predecessors, would happily have left the paradoxes of infinity to those who enjoyed such things. These practical men of science glanced at infinity, saw that it could not be tamed, and booted it out the door so that they could concentrate on the real-life questions that preoccupied them.

No sooner had they set to work than they heard a clawing at the window.

The most basic challenge in seventeenth-century science was to describe how objects move. To move is to change position. Infinity kept fighting its way into the picture because change comes in two forms. One is easy. The other would challenge and tantalize some of the most powerful thinkers the world had ever seen.

The easy form is steady change, as when a car rolls down the highway with the cruise control set at sixty miles an hour. The car is changing position, but one moment looks much like another. Now think of a rock falling off a cliff. The rock is changing position, like the car, but it is changing speed at every instant, too. That kind of *changing* change happens all around us. We see it when a population grows, or a bullet tears through the air, or an epidemic sweeps through a city. Something is changing, and the rate at which it is changing is changing, too.

Look again at the falling rock. Galileo showed that, as time passed, the rock fell faster and faster. At every instant its speed was different. But what did it mean to talk about speed at a given instant? As it turned out, that was where infinity came in. To answer even the most mundane question—how fast is a rock moving?—these seventeenth-century scientists would have to grapple with the most abstract, highfalutin question imaginable: what is the nature of infinity?

It was easy to talk about *average* speed, which posed no abstruse riddles. If a traveler in a hackney coach covered a distance of ten miles in an hour, then his average speed was plainly ten miles per hour. But what about speed not over a long interval but at a specific moment? That was trouble. What if the horses pulling the coach labored up a steep hill and then sped down the far side and then stumbled and slowed for a moment and then regained their footing and sped back up? With the coach's speed varying unpredictably, how could you possibly know its speed at a precise instant, at, for instance, the moment it passed in front of the Fox and Hounds Tavern?

The point wasn't that anyone needed to know precisely how

fast coaches traveled. For any practical question about making a journey from here to there, a rough guess would do. The coach's speed was only important as the key to a larger question: how could you devise a mathematical language that captured the ever-changing world and all its myriad moving parts? How could you see the world as God saw it?

Before you could tackle the world in general, then, it made sense to try to sort out something as familiar as a horse-drawn coach. For decades mathematicians had all tried to solve the mystery of instantaneous speed in the same way. Speed, they knew, was a measure of how much distance the coach covered in a given time. Suppose the coach happened to pass the Fox and Hounds at precisely noon. To get a rough guess of its speed at that moment, you might see how far down the road it was an hour later. If the coach had traveled eight miles between noon and one o'clock, its speed at noon was likely somewhere near eight miles per hour. But maybe not. An hour is a long while, and anything could have happened during that time. The horses might have stopped to graze the grass. They might have been stung by hornets and broken into a sprint. It would be better to guess the coach's speed at the stroke of noon by looking at how far it traveled in a shorter interval that included noon, such as from noon to 12:30. A shorter interval still, say from noon to 12:15, would be better yet. From noon to 12:01 would be even better, and from noon to one second after noon would be better than that.

Success seemed close enough to touch. To measure speed at the instant the clock struck noon, all you had to do was look at how much distance the coach covered in shorter and shorter intervals beginning at noon.

And then, with victory at hand, it flew out of reach. An instant, by definition, is briefer than the tiniest fraction of a second. How much distance did the coach cover in an instant? No distance at all, because it takes some amount of time to travel even the shortest distance. "Ten miles per hour" is a perfectly sensible speed. What could "zero distance in zero seconds" possibly mean?

OUT OF THE WHIRLPOOL

The answer began with Descartes' graphs. Since steady motion was far easier to deal with than uneven motion, scientists started there. Imagine a man trudging home from work at the end of a long day, dragging himself along at 2 miles per hour. A younger colleague might scoot along at 4 miles per hour. A runner might whiz by at 8 miles per hour.

We could chart their journeys in a table that shows how much distance they covered.

	30 MIN	60 MIN
Trudger	1 mile	2 miles
Young man	2 miles	4 miles
Runner	4 miles	8 miles

But a graph, à la Descartes, makes matters clearer. A steady pace corresponds to a straight line, as we see in the drawing below, and the faster the pace the steeper the line's slope. Slope, in other words, is a measure of speed. (*Slope* is a textbook term with a symbol-laden definition, but the technical meaning is the same as the everyday one. A line's slope is simply a measure of how quickly a situation is changing. A flat slope means no

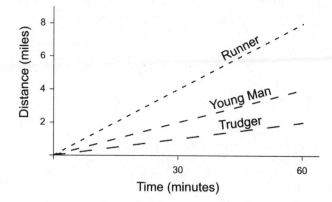

change at all; a steep slope, like a spike in blood pressure, means a fast change.)

So we can say, with the aid of our picture, precisely what it means to travel at a steady speed of 2 miles per hour (or 4, or 8). It means that, if we were to make a graph of the traveler's path, the result would be a straight line with a certain slope.

This seems simple enough, and so it is, but there is a subtle point hidden inside our tidy graph. The picture has let us dodge a vital but tricky question: What does it mean to travel at two miles per hour *if you're not traveling for a full hour?* Before Descartes came along, such questions had spawned endless confusion. But we have no need to vanish into the philosophical fog. We can nearly do without words and debates and definitions altogether. At least in the case of a traveler moving at a steady speed, we can blithely make statements like, "*At this precise instant* she's traveling at a rate of two miles *per hour.*" All this with the aid of a graph.

But suppose our task was to look at a more complicated journey than a steady march down the street. What does a graph of a cannonball's flight look like? Galileo knew that. It looks like this, as we have seen before.

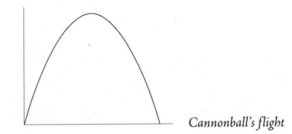

Cannonball's flight

Now our trick seems to have let us down. So long as we were dealing with graphs of straight lines, we'd found a way to talk about instantaneous speed. It was easy to talk about the slope of a straight line, because the slope was always the same. But what does it mean to talk about the slope of a curve, which by definition is *never* straight?

The question was important for two reasons. First, few changes in real life are as simple as the steady *plink plink plink* of drops from a leaky faucet. Second, if thinkers could devise a way to deal with complicated change of *one* sort, then presumably they could deal with complicated changes of *many* sorts. Mathematics is so powerful—and so difficult for us to learn— because it is a universal tool. We balk at algebra, for instance, because those inscrutable *x*'s are so off-putting. But algebra is useful precisely because it allows us to fill in the blanks in countless different ways.

A mathematics of change dangled the same promise. Planets and comets speeding across the heavens, populations growing and shrinking, bank accounts swelling, divers plummeting, snowbanks melting, all would yield up their mysteries. Questions that asked when a given change would reach its high point or its low—what angle should a cannon be tilted at to shoot the farthest? when will a growing population level off? what is the ideal shape for the arch of a bridge?—could be answered quickly and definitively.

This was a gleaming prize. But how to win it?

The riddle at the heart of the mystery of motion was the question of speed at a given instant. What did it mean? How could you keep from drowning in the whirlpool of "zero distance in zero seconds"?

Answering that question meant learning how to focus on infinitesimally brief stretches of time. The first step was to see that Zeno was not so unnerving as he had seemed. Take his argument that it would take forever to cross a room because it would take a certain amount of time to cross to the halfway point, and then more time to cross half the remaining distance, and so on.

Zeno's paradox. If it takes 1 second to walk to the middle of a room, and ½ second more to walk half the rest of the way, and ¼ second more to walk halfway again, and so on, then it takes an infinitely long time to cross the room.

In essence Zeno's argument is a claim about infinity. It seems common sense to say that if you add up numbers forever, and if each number you add is bigger than zero, then eventually the sum is infinite. If you piled up blocks *forever*, wouldn't the stack eventually reach the ceiling, no matter how vast a room you started in?

Well, no, actually, not necessarily.

It all depends on the size of each new block that you added to the stack. If all the blocks were the same size, then the tower *would* eventually reach the ceiling, the moon, the stars. And even if each new block were thinner than its predecessor, a tower might still grow forever.* But it might not, if you picked the sizes of the blocks just so.

In modern terms, Zeno's paradox amounts to saying that if you add up $1 + \frac{1}{2} + \frac{1}{4} + \frac{1}{8} + \frac{1}{16} + \ldots$ the total is infinite. Zeno never framed it that way. He focused not on a specific chain of fractions like this one, but on a general argument about what had to be true for *any* endless list of numbers whatsoever.

But Zeno was wrong. If the sum were infinite, as he believed, then it would be bigger than any number you can think of— bigger than 100, bigger than 100,000, and so on. But Zeno's sum does *not* exceed any number you can name. On the contrary, the sum is the perfectly ordinary number 2.

In a minute, we'll see why that is. But think how surprising this result is. Suppose you took a block one inch high, and put a ½-inch-thick block on top of it, and then a ¼-inch-thick one on top of that, and so on. If you added new blocks forever, one a second, through your lifetime and your children's lifetimes and the universe's lifetime, even so the tower would never reach above a two-year-old's ankles.

---

* If the first block were 1 inch thick, the next ½ inch, then ⅓, ¼, ⅕, and so on, the tower *would* climb infinitely high (although it would rise excruciatingly slowly).

Part Three

INTO THE LIGHT

ALL MEN ARE CREATED EQUAL

The taming of infinity represents another of those break-throughs where a once-baffling abstraction, like "zero" or "negative five," comes to seem simple in hindsight. The key was to stay relentlessly down-to-earth and never to venture into such murky territory as "the nature of infinity."

The abstraction that would save the day was the notion of a "limit." The mathematical sense is close to the everyday one. In one of the Lincoln-Douglas debates, Abraham Lincoln asked his listeners why the Declaration of Independence asserted that "all men are created equal." Not because the founders believed that all men had already attained equality, Lincoln said. *That* was an "obvious untruth." The founders' point, Lincoln declared, was that equality for all was a goal that should be "constantly looked to, constantly labored for, and even though never perfectly attained, constantly approximated."

In the same sense, a mathematical limit is a goal, a target that a sequence of numbers comes ever closer to. The sequence doesn't have to *reach* the limit, but it does have to get nearer and nearer. The limit of the sequence 1, .1, .01, .001, .0001, . . . is the number 0, even though the sequence never gets there. Similarly, the limit of ½, ¾, $^5/_6$, $^7/_8$, $^9/_{10}$, $^{10}/_{11}$, ... is the number 1, also never attained. The sequence 1, 2, 1, 2, 1, 2, ... does not have a limit, because it hops back and forth forever and never homes in on a target.*

Zeno cast his paradox in the form of a story about a journey across a room. In the 1500s and 1600s a few intrepid mathematicians reframed his tale as a statement about numbers. From that perspective, the question was whether or not 1 + ½ + ¼ + ⅛ + $^1/_{16}$ + ... added up to infinity. Zeno's answer was "yes," because the numbers go on forever and each contributes something to the sum. But when mathematicians turned from Zeno's words to their numbers and began adding, they found something odd. They began with 1 + ½. That made 1½. Nothing dire there. How about 1 + ½ + ¼? That came to 1¾. Still okay. 1 + ½ + ¼ + ⅛? That was 1⅞. They added more and more terms and never ran into trouble. The running total continued to grow, but it became ever clearer that the number 2 represented a kind of boundary. You could draw arbitrarily close to that boundary—within one-thousandth or one-billionth or even closer—but certainly you could never break through in the way that a runner breaks the tape at the finish line.

For the practical-minded scientists of the seventeenth century, this meant the end of Zeno. In the battle with infinity, they declared victory. Zeno had maintained that if it took one second to reach the middle of a room, it would take *forever* to

---

* A sequence *may* attain its goal. The sequence 1, 1, 1, ... has the number 1 as its limit. But a "typical" sequence draws ever nearer to its goal without actually touching it. The sequence .9, .99, .999, ... never reaches its limit, which is the number 1.

cross to the other side. Not so, said the new mathematicians. It would take two seconds.

Why did that strike them as so momentous? Because when they had taken up the question they truly wanted to answer—what does *instantaneous speed* mean?—they had run head-on into Zeno's paradox. They had wanted to know a hackney coach's speed at the instant of noon and had found themselves ensnarled in an infinite regress of questions of the form, *what was the coach's speed between 12:00 and one minute after? between 12:00 and 30 seconds after? between 12:00 and 15 seconds after? between 12:00 and . . . ?*

This was the seventeenth-century's counterpart of phone-menu hell ("if your call is about billing, press 1"), and early scientists groaned in despair because the questions continued endlessly, and escape seemed impossible. But now their victory over Zeno gave them hope. Yes, the questions about the coach *did* go on forever. But suppose you looked at the coach's speed in briefer and briefer intervals and found that that sequence of speeds homed in on a limit?

Then your troubles would be over. That limit would be a number—a definite, perfectly ordinary number. *That* was what "instantaneous speed" meant. Nothing to it. But the greatest mathematicians of antiquity, and all their descendants for another fifteen centuries, had failed to see it.

This was not quite calculus, but it was a giant step toward it. In essence, calculus would be a mathematical microscope, a tool that let you pin motion down and scrutinize it tip to toe. Some moments were more important than others—the arrow's height at the instant it reached its peak, the cannonball's speed at the instant it smashed into a city's wall, a comet's speed when it rounded the sun—and, with calculus's help, you could fix those particular moments to a slide and study them close-up.

Or so the newly optimistic mathematicians presumed. But when they grabbed the microscope, they found that no matter how they twisted and tweaked its knobs they simply could not bring the image into focus. The problem, they soon saw, was that everything hinged on the notion of limits, and limits weren't as straightforward as they had thought.

As with all other abstractions, the problem was trying to wrestle with a phantom. What did it mean, precisely and quantitatively, for a sequence of numbers to come very close to a limit? "The planet Mars comes close to the Earth when it is 50 million miles away," one modern mathematician observes. "On the other hand, a bullet comes close to a person if it gets within a few inches of him." How close is close?

Even Isaac Newton and Gottfried Leibniz, the boldest thinkers of their age and the leaders of the assault on infinity, found themselves tangled up in confusion and contradiction. For one thing, infinity seemed to come in a disarming variety of forms. In ordinary usage, *infinity* conjured up thoughts of boundless immensity. Now, though, in all this talk of speed at a given instant, it seemed vital to sort out the meaning of "infinitely small" lengths and "infinitely brief" stretches of time, as well.

Worse still, the tiny distances and the tiny intervals of time were all mingled together. Speed means distance divided by time. That was not a problem when you were dealing with large, familiar units like miles and hours. But how could you keep your eyes from blurring when it came to dividing ever-shorter distances by ever-briefer time spans?

No one could think how to classify these vanishingly small times and lengths. Leibniz talked of "infinitesimals," which were by definition "the smallest possible numbers," but that definition raised as many questions as it answered. How could a number be smaller than *every* fraction? Perhaps infinitesimals were real but too small to see, like the microscopic creatures Leeuwenhoek had recently discovered? As tiny as they were, infinitesimals were bigger than 0. Except sometimes, when they weren't.

Leibniz tried to explain, but he only made matters worse. "By . . . infinitely small, we understand something . . . indefinitely small, so that each conducts itself as a sort of class, and not merely as the last thing of a class. If anyone wishes to understand these [the infinitely small] as the ultimate things . . . it can be done." This was, two of Leibniz's disciples acknowledged, "an enigma rather than an explication." Newton spoke instead of "the ultimate ratio of evanescent quantities," which was perhaps clear to him but baffling to almost everyone else. "In mathematics the minutest errors are not to be neglected," he insisted in one breath, and in the next he pointed out that these tiny crumbs of numbers were so close to 0 that they could safely be ignored.

Amazingly, things mostly worked out, much as earlier generations had found that things mostly worked out when they manipulated what were then newfangled and still mysterious negative numbers. In the case of calculus, a seemingly mystical abracadabra yielded utterly down-to-earth, hardheaded results about such questions as how far cannonballs would travel and how much damage they would do when they landed. The very name *calculus* served as a testimonial to the practical value of this new art; *calculus* is the Latin word for "pebble," a reference to the heaps of stones once used as a calculating aid in addition and multiplication.

Skeptics contended that any correct results must have been

due to happy accidents in which multiple errors canceled themselves out. ("For science it cannot be called," one critic later charged, "when you proceed blindfold and arrive at the Truth not knowing how or by what means.") But so long as the slapdash new techniques kept churning out answers to questions that had always lain out of reach, no one spent much time worrying about rigor. Leibniz, boundlessly optimistic in personality as well as in his philosophical views, argued explicitly that this gift horse should be saddled and ridden, not inspected. It would all work out.

The muddle would last until the 1800s. Only then would a new generation of mathematicians find a way to replace vague intuitions with clear definitions. (The breakthrough was finding a way to define "limits" while banishing all talk of infinitely small numbers.) In all the intervening years mathematicians and scientists had rejoiced in a bounty they did not understand. Instead they followed the advice of Jean d'Alembert, a French mathematician who lived a century after Newton and Leibniz but during the era when the underpinnings of calculus were still cloaked in mystery.

"Persist," d'Alembert advised, "and faith will come to you."

Chapter Thirty-Eight

THE MIRACLE YEARS

Both Isaac Newton and Gottfried Leibniz had egos as colossal as their intellects. In the hunt for calculus, each man saw himself as a lone adventurer in unexplored territory. And then, unbeknownst to one another, each gained the prize he sought. Each saw his triumph not as that of a runner bursting past a pack of rivals but as that of a solo mountain climber. They had won their way to a summit, moreover, that no one else even knew existed. Or so they both believed.

Imagine, then, the exultation that each man felt when he planted his flag in the ice and gazed out at the panorama before him, a landscape that extended as far as the eye could see. Picture, too, the pride and satisfaction that came with lone ownership of this vast and beckoning domain. And then imagine the morning when that proprietary delight gave way to shock and horror. Imagine the first glimpse of a puff of smoke in the distance—*fog, surely, for how could anyone have built a fire in this emptiness?*—and then, soon after, the unmistakable sight of someone else's footprints in the snow.

Newton had been the first to learn how to pin down the mysterious infinitesimals that held the key to explaining motion. He kept his discoveries secret from all but a tiny circle for three decades. Victimized by his own temperament—Newton was always torn between indignation at seeing anyone else get credit for work he had done first and fury at the thought of announcing his findings and thereby exposing himself to critics—he might have hesitated forever. As it was, his delay in staking his claim led to one of the bitterest feuds in the history of science.

Newton made his mathematical breakthroughs (and others just as important) in a fever of creativity that historians would later call the "miracle years." He spent eighteen months between 1665 and 1667 at his mother's farm, hiding from the plague that had shut Cambridge down. Newton was twenty-two when he returned home, undistinguished, unknown, and alone.

Intellectually, though, he was not entirely on his own. Calculus was in the air, and such eminent mathematicians as Fermat, Pascal, and Descartes had made considerable advances toward it. Newton had attended mathematical lectures at Cambridge; he had bought and borrowed a few textbooks; he had studied Descartes' newfangled geometry with diligence.

What sparked his mathematical interest in the first place he never said. We can, however, pin down the time and place. Every August, Cambridge played host to an enormous outdoor market called Stourbridge Fair. In row upon row of tents and booths, merchants and hawkers sold clothes, dishes, toys, furniture, books, jewelry, beer, ale, and, in the horrified words of John Bunyan, "lusts, pleasures, and delights of all sorts." Newton steered well clear of the swarms of whores, jugglers, and con men. (He had given a good deal of thought to temptation, sexual temptation above all, and had fashioned a strategy. "The way to chastity is not to struggle directly with incontinent thoughts," he wrote in an essay on monasteries and the early church, "but to avert ye thoughts by some employment, or by reading, or meditating on other things.")

Newton made two purchases. They seemed innocuous, but they would revolutionize the intellectual world. "In '63 [Newton] being at Stourbridge fair bought a book of astrology to see what there was in it," according to a young admirer who had the story from Newton himself. Perhaps in the same year—scholars have not settled the matter—he bought a trinket, a glass prism. Children liked to play with prisms because it was pretty to see how they caught the light.

The astrology book had no significance in itself, but it helped change history. Newton "read it 'til he came to a figure of the heavens which he could not understand for want of being acquainted with trigonometry," he recalled many years later. "Bought a book of trigonometry, but was not able to understand the demonstrations. Got Euclid to fit himself for understanding the ground of trigonometry."

At that point Newton's backtracking came to an end. To his relief, he found that Euclid was no challenge. "Read only the titles of the propositions," he would recall, "which he found so easy to understand that he wondered how anybody would amuse themselves to write any demonstrations of them."

Newton turned from Euclid's classical geometry to Descartes' recent recasting of the entire subject. This was not so easy. He made it through two or three pages of Descartes but then lost his way. He started over and this time managed to understand three or four pages. He slogged along in this fashion, inching his way forward until he lost his bearings and then doubling back to the beginning "& continued so doing till he made himself Master of the whole without having the least light or instruction from any body." Every aspiring mathematician knows the frustration of spending entire days staring at a single page in a textbook, or even a single line, waiting for insight to dawn. It is heartening to see one of the greatest of all mathematicians in almost the same plight.

Newton's pride in finally mastering Descartes' *Geometry* had two aspects, and both were typical of him. He had accomplished a great deal, and he had done it without a word of guidance "from any body." And he had only begun. To this point he had studied work that others had already done. From here on, he would be advancing into unexplored territory. In early 1665, less than two years from the day he had picked up the astrology booklet, he recorded his first mathematical discovery. He proved what is now called the binomial theorem, to this day one of the essential results in all of mathematics.* This was the opening salvo of the "miracle years."

Newton's summary of what came next remains startling three and a half centuries later. Even those unfamiliar with the vocabulary cannot miss the rat-tat-tat pacing of discoveries that spilled out almost too quickly to list. "The same year in May I found the method of Tangents . . . & in November had the direct method of fluxions & the next year in January had the Theory of Colours & in May following I had entrance into ye inverse method of fluxions. And the same year I began to think of gravity extending to ye orb of the Moon. . . ."

Over the course of eighteen months, that is, Newton first

---

* Gilbert and Sullivan's Major-General knew it well, along with much else. "About binomial theorem I'm teeming with a lot o' news / With many cheerful facts about the square of the hypotenuse."

invented a great chunk of calculus, everything to do with what is now called differentiation. Then he briefly put mathematics aside and turned to physics. Taking up his Stourbridge Fair prisms (he had bought a second one) and shutting up his room except for a pinhole that admitted a shaft of sunlight, he discovered the nature of light. Then he turned back to calculus. The subject falls naturally into two halves, although that is by no means evident early on. In early 1665 Newton had invented and then investigated the first half; now he knocked off the other half, this time inventing the techniques now known as integration. Then he proved that the two halves, which looked completely different, were in fact intimately related and could be used in tandem in hugely powerful ways. Then he began thinking about the nature of gravity. "All this," he wrote, "was in the two plague years of 1665–1666. For in those days I was in the prime of my age for invention & minded Mathematicks & Philosophy more than at any time since."

Newton was indeed in his prime at twenty-three, for mathematics and physics are games for the young. Einstein was twenty-six when he came up with the special theory of relativity, Heisenberg twenty-five when he formulated the uncertainty principle, Niels Bohr twenty-eight when he proposed a revolutionary model of the atom. "If you haven't done outstanding work in mathematics by 30, you never will," says Ronald Graham, one of today's best-regarded mathematicians.

The greats flare up early, like athletes, and they burn out just as quickly. Paul Dirac, a physicist who won his Nobel Prize for work he did at twenty-six, made the point with wry bleakness, in verse. (He wrote his poem while still in his twenties.)

Age is, of course, a fever chill
that every physicist must fear.
He's better dead than living still
when once he's past his thirtieth year.

In the most abstract fields—music, mathematics, physics, even chess—the young thrive. Child prodigies are not quite common, but they turn up regularly. Perhaps it makes sense that if a Mozart or a Bobby Fischer were to appear anywhere, it would be in a self-contained field that does not require insight into the quirks of human psychology. We are unlikely ever to meet a twelve-year-old Tolstoy.

But that is only part of the story. Penetrating to the heart of abstract fields seems to demand a degree of intellectual fire-power, an intensity of focus and stamina, that only the young can muster. For the greats, these truly are miracle years. "I know that when I was in my late teens and early twenties the world was just a Roman candle—rockets all the time," recalled I. I. Rabi, another Nobel Prize–winning physicist. "You lose that sort of thing as time goes on. . . . Physics is an otherworld thing. It requires a taste for things unseen, even unheard of—a high degree of abstraction. . . . These faculties die off somehow when you grow up."

Nerve and brashness are as vital as brainpower. A novice sets out to change the world, confident that he can find what has eluded every other seeker. The expert knows all the reasons why the quest is impossible. The result is that the young make the breakthroughs. The pattern is different in the arts. "Look at a composer or a writer—one can divide his work into early, middle, and late, and the late work is always better, more mature," observed Subrahmanyan Chandrasekhar, the astrophysicist who won a Nobel Prize for his work on black holes (and worked into his eighties). Even so, he declared in his old age, "For scientists, the early work is always better."

At age thirty-five or forty, when a politician would still count as a fresh face, when doctors in certain specialties might only recently have completed their training, mathematicians and physicists know they have probably passed their peak. In the arts, talent often crests at around forty. Michelangelo completed the ceiling of the Sistine Chapel at thirty-seven; Beethoven finished his Fifth Symphony at thirty-seven; Tolstoy published *War and Peace* at forty-one; Shakespeare wrote *King Lear* at forty-two. But the list of artists who continued to produce masterpieces decades later than that—Monet, Cervantes, Titian, Picasso, Verdi—is long.

Science and mathematics have no such roster. In the end, the work simply becomes too difficult. Newton would make great advances in mathematics after his miracle years, but he would never again match the creative fervor of that first outburst. Looking back at his career in his old age, he remarked that "no old Men (excepting Dr. Wallis)"—this was Newton's eminent contemporary John Wallis—"love Mathematicks."

From his earliest youth, Newton had seen himself as different from others, set apart and meant for special things. He read great significance into his birth on Christmas Day, his lack of a father, and his seemingly miraculous survival in infancy. The depth and sincerity of his religious faith are beyond question, and so was his belief that God had set him apart and whispered His secrets into his ear. Others had studied the prophecies in the Bible just as he had, Newton noted, but they had met only "difficulty & ill success." He was unsurprised. Understanding was reserved for "a remnant, a few scattered persons which God hath chosen." Guess who.

He took the Latin form of his name, Isaacus Nevtonus, and found in it an anagram, *Ieova sanctus unus*, or *the one holy Jehovah*. He drew attention to the passage in Isaiah where God promises the righteous that "I will give thee the treasures of darkness, and hidden riches of secret places."

By the end of the miracle years, Newton found himself awash in hidden riches. He knew more mathematics than anyone else in the world (and therefore more than anyone who had ever lived). No one even suspected. "The fact that he was unknown does not alter the other fact that the young man not yet twenty-four, without benefit of formal instruction, had become the leading mathematician of Europe," wrote Richard Westfall, Newton's preeminent biographer. "And the only one who really mattered, Newton himself, understood his position clearly enough. He had studied the acknowledged masters. He knew the limits they could not surpass. He had outstripped them all, and by far."

Newton had always *felt* himself isolated from others. Now at twenty-three, wrote Westfall, he finally had objective proof that he was not like other men. "In 1665, as he realized the full extent of his achievement in mathematics, Newton must have felt the burden of genius settle upon him, the terrible burden which he would have to carry in the isolation it imposed for more than sixty years."

Chapter Thirty-Nine

ALL MYSTERY BANISHED

Isaac Newton believed that he had been tapped by God to decipher the workings of the universe. Gottfried Leibniz thought that Newton had set his sights too low. Leibniz shared Newton's yearning to find nature's mathematical structure, which in their era meant almost inevitably that both men would mount an assault on calculus, but in Leibniz's view mathematics was only one piece in a much larger puzzle.

Leibniz was perhaps the last man who thought it was possible to know everything. The universe was perfectly rational, he believed, and its every feature had a purpose. With enough attention you could explain it all, just as you could deduce the function of every spoke and spring in a carriage.

For Leibniz, one of the greatest philosophers of the age, this was more than a demonstration of almost pathological optimism (though it was that, too). More important, Leibniz's faith was a matter of philosophical conviction. The universe *had* to make perfect sense because it had been created by an infinitely wise, infinitely rational God. To a powerful enough intellect, every true observation about the world would be self-evident, just as every true statement in geometry would immediately be obvious. In all such cases, the conclusion was built in from the start, as in the statement "all bachelors are unmarried." We humans might not be clever enough to see through the undergrowth that obscures the world, but to God every truth shines bright and clear.

In fact, though, Leibniz felt certain that God had designed the world so that we *can* understand it. Newton took a more cautious stand. Humans could read the mind of God, he believed, but perhaps not all of it. "I don't know what I may seem to the world," Newton famously declared in his old age, though he knew perfectly well, "but, as to myself, I seem to have been only like a boy playing on the seashore, and diverting myself in now and then finding a smoother pebble or a prettier shell than ordinary, whilst the great ocean of truth lay all undiscovered before me."

Newton's point was not simply that some questions had yet to be answered. Some questions might not *have* answers, or at least not answers we can grasp. Why had God chosen to create something rather than nothing? Why had He made the sun just the size it is? Newton believed that such mysteries might lie beyond human comprehension. Certainly they were outside the range of scientific inquiry. "As a blind man has no idea of colors," Newton wrote, "so have we no idea of the manner by which the all-wise God perceives and understands all things."

Leibniz accepted no such bounds. God, he famously declared, had created the best of all possible worlds. This was not an assumption, in Leibniz's view, but a deduction. God was by definition all-powerful and all-knowing, so it followed at once that the world could not have been better designed. (Even for one of the ablest of all philosophers, this made for an impossible tangle. If logic compelled God to create the very world we find ourselves in, didn't that mean that He had no choice in the matter? But surely to be God meant to have *infinite* choice?)

Voltaire would later take endless delight, in *Candide*, in pummeling Leibniz. On *Candide*'s very first page, we meet Leibniz's stand-in, Dr. Pangloss, the greatest philosopher in the world. Pangloss's specialty is "metaphysico-theologo-cosmolonigology." The world, Pangloss explains contentedly, has been made expressly for our benefit. "The nose is formed for spectacles, therefore we wear spectacles. . . . Pigs were made to be eaten, therefore we eat pork all the year round."

Pangloss and the hero of the novel, a naïve young man named Candide, spend the book beset by calamity—Voltaire cheerily throws in an earthquake, a bout of syphilis, a stint as a galley slave, for starters. Bloodied and battered though both men may be, Pangloss pops up from every crisis as undaunted as a jack-in-the-box, pointing out once more that this is the best of all possible worlds.

This was great fun—Voltaire was an immensely popular writer, and *Candide* was his most popular work—but it was a bit misleading. Leibniz knew perfectly well that the world abounded in horrors. (He had been born during the Thirty Years' War.) His point was not that all was sunshine, but that no better alternative was possible. God had considered every conceivable universe before settling on this one. Other universes might have been good, but ours is better. God could, for instance, have made humans only as intelligent as dogs. That might have made for a happier world, but happiness is not the only virtue. In a world of poodles and Great Danes, who would paint pictures and write symphonies?

Or God might have built us so that we always chose to do good rather than evil. In such a world, we would all be kind, but we would all be automatons. In His wisdom, God had decided against it. A world with sin was better than a world without choice. Not perfect, in other words, but better than any possible alternative. It was this complacency that infuriated Voltaire. He raged against Leibniz not because Leibniz was blind to the world's miseries but because he so easily reconciled himself to them.

But Leibniz's God was as rational as he was. For every conceivable world, He totted up the pros and cons and then subtracted the one from the other to compute a final grade. (It is perhaps no surprise that Leibniz invented calculus; in searching for the world that would receive the highest possible score, God was essentially solving a calculus problem.) Since God had necessarily created the best of all possible worlds, Leibniz went on, we can deduce its properties by pure thought. The best possible world was the one that placed the highest value on the pursuit of intellectual pleasure—here the philosopher showed his hand—and the greatest of all intellectual pleasures was finding order in apparent disorder. It was certain, therefore, that God meant for us to solve all the world's riddles. Leibniz was "perhaps the most resolute champion of rationalism who ever appeared in the history of philosophy," in the words of the philosopher Ernst Cassirer. "For Leibniz there . . . is nothing in heaven or on earth, no mystery in religion, no secret in nature, which can defy the power and effort of reason."

Surely, then, Leibniz could solve the problem of describing the natural world in the language of mathematics.

Chapter Forty

TALKING DOGS AND UNSUSPECTED POWERS

Leibniz gave the impression that he intended to pursue every one of nature's secrets himself. "In the century of Kepler, Galileo, Descartes, Pascal, and Newton," one historian wrote, "the most versatile genius of all was Gottfried Wilhelm Leibniz." The grandest topics intrigued him, and so did the humblest. Leibniz invented a new kind of nail, with ridged sides to keep it from working free. He traveled to see a talking dog and reported to the French Academy that it had "an aptitude that was hard to find in another dog." (The wondrous beast could pronounce the French words for tea, coffee, and chocolate, and some two dozen more.)

He drew up detailed plans for "a museum of everything that could be imagined," roughly a cross between a science exhibition and a Ripley's Believe It or Not museum. It would feature clowns and fireworks, races between mechanical horses, rope dancers, fire eaters, musical instruments that played by themselves, gambling halls (to bring in money), inventions, an anatomical theater, transfusions, telescopes, demonstrations of how the human voice could shatter a drinking glass or how light reflected from a mirror could ignite a fire.

Leibniz's energy and curiosity never flagged, but he could scarcely keep up with all the ideas careening around his head. "I have so much that is new in mathematics, so many thoughts in philosophy, so numerous literary observations of other kinds, which I do not wish to lose, that I am often at a loss what to do first," he lamented.

Many of these ventures consumed years, partly because they were so ambitious, partly because Leibniz tackled everything at once. He continued to work on his calculating machine, for example, and on devising a symbolic language that would allow disputes in ethics and philosophy to be solved like problems in algebra. "If controversies were to arise, there would be no more need of disputing between two philosophers than between two accountants. For it would suffice to take their pencils in their hands, to sit down to their slates, and to say to each other (with a friend as witness, if they liked): 'Let us calculate.'"

Leibniz wrote endlessly, at high speed, often while bumping along the road in a coach. Today a diligent team of editors is laboring to turn well over one hundred thousand manuscript pages into a Collected Works, but they do not expect to complete the project in their lifetimes. Volume 4, to choose an example at random, comes under the heading of "Philosophical Writings," and consists of three "books." Each book contains over a thousand pages. The editors envision sixty such volumes.

Thinkers who take on the whole world, as Leibniz did, are out of fashion today. Even in his own era, he was a hard man to get the measure of. Astonishingly brilliant, jaw-droppingly vain, charming, overbearing, a visionary one minute and a self-deluded dreamer the next, he was plainly a lot of work. Not everyone was inclined to make the effort. Still, in Bertrand Russell's words, "Leibniz was one of the supreme intellects of all time." If anything, his reputation among scientists and mathematicians has grown through the centuries, as ideas of his that once seemed simply baffling have come into focus.

More than three hundred years ago, for instance, Leibniz envisioned the digital computer. He had discovered the binary language of 0s and 1s now familiar to every computer programmer,* and, more remarkably, he had imagined how this two-letter alphabet could be used to write instructions for an all-purpose reasoning machine.

The computer that Leibniz had in mind relied not on electrical signals—this was almost a century before Benjamin Franklin would stand outdoors with a kite in a lightning storm—but on marbles tumbling down chutes in a kind of pinball machine. "A container shall be provided with holes in such a way that they can be opened and closed," Leibniz wrote. "They are to be open at those places that correspond to a 1 and remain closed at those that correspond to a 0. Through the opened gates small cubes or marbles are to fall into tracks, through the others nothing."

Leibniz was born in Germany, but he spent his glory years in the glittering Paris of Louis XIV, when the Sun King had just begun building Versailles and emptying the royal treasury. Leibniz arrived in Paris in 1672, at age twenty-six, a dapper young diplomat sporting a long wig. The dark curls and silk stockings were standard fare, but the torrent of words that spilled forth

---

* Unbeknownst to Leibniz, the English mathematician and astronomer Thomas Harriot had been the first to discuss binary numbers, decades before. But Harriot never published any of his work, and his papers went unseen until the late 1700s. It turns out that Harriot had recorded a number of other firsts as well; Harriot turned a telescope to the sky a few weeks before Galileo did.

from the new arrival dazed his listeners. Leibniz had come to Paris with characteristically bold plans. Germany dreaded an invasion by the French, who had grand territorial ambitions. Leibniz's mission was to convince Louis XIV that an incursion into Germany would do him little good. What he ought to do instead, what would prove a triumph worthy of so illustrious a monarch, was to conquer Egypt.

In four years Leibniz never managed to win an audience with the king. (France, as everyone had feared, spent the next several decades embroiling Europe in one war after another.) Leibniz spent his time productively nonetheless, somehow combining an endless series of visits with one count or duke or bishop after another with the deepest investigations into science and mathematics.

Leibniz's conquest of mathematics came as a surprise. Unlike nearly all the other great figures in the field, he came to it late. Leibniz's academic training had centered on law and diplomacy. In those fields, as well as philosophy and history and a dozen others, he knew everything. But at twenty-six, one historian writes, Leibniz's knowledge of mathematics was "deplorable."

He would remedy that. In Paris he set to work under the guidance of some leading mathematicians, notably the brilliant Dutch scientist Christiaan Huygens. For the most part, though, he taught himself. He took up classic works, like Euclid, and recent ones, like Pascal and Descartes, and dipped in and out at random like a library patron flipping through the books on the "new arrivals" shelf. Even Newton had found that newfangled doctrines like Descartes' geometry slowed him to a crawl. Not Leibniz. "I read [mathematics] almost as one reads tales of romance," he boasted.

He read voraciously and competitively. These were difficult, compact works by brilliant men writing for a tiny audience of peers, not textbooks meant for students, and Leibniz measured himself against the top figures in this new field. "It seemed to me," he wrote shortly after beginning his crash course, "I do not know by what rash confidence in my own ability, that I might become the equal of these if I so desired." The time had come to stop reading about what other people had done and to make discoveries of his own.

By now it was 1675. Leibniz was thirty but still, at that advanced mathematical age, at the peak of his powers. The riddle that taunted every mathematician was the infinitesimal, the key to understanding motion at a given instant. Nearly a decade before, Newton had solved the mystery and invented what is now called calculus. He had told almost no one, preferring to wrap that secret knowledge around himself like a warm cloak. Now, unaware of what Newton had already done, Leibniz set out after the same prize.

In the course of one astonishing year—a miracle year of his own—he found it. Newton had kept his discovery to himself, because of his hatred of controversy and because the security of his professorship at Cambridge meant he did not have to scramble for recognition. Leibniz did not publish an account of his discovery of calculus for nine years, but his silence is harder to explain. Leibniz never had a safe position like Newton's. Throughout his long career, he was dependent on the whims of his royal patrons, forever trapped in the role of an intellectual court jester. That might have made him *more* eager to publish, anything to make his status less precarious, but it did not.

The reasons for his delay have disappeared into a biographical black hole. Leibniz wrote endlessly on every conceivable topic— his correspondence alone consisted of fifteen thousand letters, many of them more essays than notes—but he remained silent on the question of his long hesitation. Scholars can only fill the void with guesses.

Perhaps he was gun-shy as a result of a fiasco at the very beginning of his mathematical career. On his first trip to England, in 1672, Leibniz had met several prominent mathematicians (but not Newton) and happily rattled on about his discoveries. The bragging was innocent, but Leibniz was such a mathematical novice that he talked himself into trouble. At an elegant dinner party in London, presided over by Robert Boyle, Leibniz claimed as his own a result (involving the sum of a certain infinitely long sequence of fractions) that was in fact well-known. Another guest set him straight. In time the episode blew over. Still, Leibniz may have decided to make sure that he stood on firm ground before he announced far bolder mathematical claims.

Or perhaps he decided that formal publication was beside the point because the audience he needed to reach had already learned of his achievement through informal channels—rumors and letters. Or the task of developing a full-fledged theory, as opposed to a collection of techniques for special cases, may have proved unexpectedly difficult. Or Leibniz may have judged that he needed to make a bigger splash—from an impossible-to-miss invention like the telescope or from some diplomatic coup— than any mathematical discovery could provide.

Eventually, in 1684, Leibniz told the world what he had discovered. By then he and Newton had exchanged friendly but guarded letters discussing mathematics in detail but tiptoeing around the whole subject of calculus. (Rather than tell Leibniz directly what he had found, Newton concealed his most important discoveries in two encrypted messages. One read,

"6accdae13eff7i319n4o4qrr4s8t12ux.") In the published article announcing his discovery of calculus, Leibniz made no mention of Newton or any of his other predecessors.

In the case of Newton, at least, that oversight was all but inevitable, since Leibniz had no way of knowing what Newton had found. A perfect alibi, one might think, but it proved anything but. Leibniz's "oversight" was destined to poison the last decades of his life.

THE WORLD IN CLOSE-UP

Newton and Leibniz framed their discoveries in different vocabulary, but both had found the same thing. The challenge confronting both men was finding a way to stop time in its tracks. Their solution, hundreds of years before the birth of photography, was essentially to imagine the movie camera. They pictured the world not as the continuous, flowing panorama we see but as a series of still photos, each one barely different from those before and after it, and all the frames flashing before the eye too quickly to register as static images.

But how could you be sure that, no matter what moment you wanted to scrutinize, there happened to be a sharply focused image on hand? It seemed clear that the briefer the interval between successive still photos, the better. The problem was finding a stopping place—if sixty-four frames a second was good, wouldn't 128 be better? Or 1,000, or 100,000?

Think of Galileo near the top of the Leaning Tower, a bit winded from the long climb. He extends an arm out into space, opens his fingers, and releases the rock he has lugged up all this way. It falls faster and faster—in each successive second, that is, it covers more distance than it did the second before—as the figures in this table show. (As we have seen, it was no easy matter to make such measurements without clocks or cameras, which was why Galileo ended up working with ramps rather than towers.)

Time (seconds)	Distance (feet)
1	16
2	64
3	144

Galileo found that rocks fall according to a precise rule that can be expressed in symbols. Scientists write the rule as $d = 16\,t^2$, where d stands for distance and t stands for time. In one second, a rock falls a distance of 16×1 feet, or 16 feet. In two seconds, it falls a distance of 16×4 feet, or 64 feet; in three seconds, 16×9 feet, or 144 feet.

The table can be converted to a graph, and, as usual, a picture helps reveal what the numbers only imply. (And a picture, unlike a table, shows the rock's position at *every* moment rather than at a select few.) The horizontal axis depicts time, the verti-

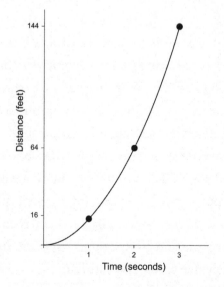

The graph shows how far a rock dropped from a height falls in t seconds. The rock obeys the rule $d = 16\,t^2$.

cal axis distance. The drawing shows the distance the rock has fallen at a given time. At the moment that Galileo uncurled his fingers and released the rock (in other words, at $t = 0$), the rock has fallen 0 feet. At 1 second it has fallen 16 feet; in 2 seconds, 64 feet; and so on.

The transition from rock to table to graph is one of increasing abstraction, and early mathematicians had a hard time keeping their bearings. Few things could be more tangible than a rock. When Galileo let it go, anyone who happened to be passing by could see it. The table took that ordinary event—a stone whooshing toward the ground—and transformed it into a list of numbers. The graph represented still another move away from everyday reality. It shows a curve that represents the rock's distance from Galileo's hand, but the curve in the graph only matches the real-life descent of the falling rock in a subtle way. The actual rock dropped in a straight line. The graph shows a curve, a parabola. Worse yet, the rock fell *down*, while the parabola headed *up*. To "see" the rock falling in the way that the graph depicts required a far more laborious and roundabout process than merely looking at a rock.

And yet it was this second way of looking at a rock's fall, this unnatural way, that held the key to nature's secrets. For it was this curve that showed Newton and Leibniz how to seize time in their fists and hold it still. We saw in chapter 36 that in a graph that shows distance measured against time, a straight line corresponds to a steady speed. (The greater the speed, the steeper the slope of that line, because a steeper slope indicates more distance covered in a given amount of time.) But in the graph of a falling rock, we have a curve, not a straight line. How can we talk about the rock's speed? In particular, how can we find its speed at a specific instant, at, for instance, precisely one second into its fall?

We could do it, Newton and Leibniz explained, if we could find the slope of the curve at precisely the one-second point. Which they proceeded to do. The idea was to look at the curve in extraordinary close-up. If you look closely enough, a curve looks like a straight line. (A jogger on a huge circular track would feel as if she were running in a straight line. Only a bird's-eye view would reveal the track's true shape.) And although curves are hard to work with, straight lines are easy.

First they froze time by selecting a single frame from nature's ongoing movie. (Newton and Leibniz worked in parallel, unaware of one another, as we have seen, but they independently hit on the same strategy.) Second, they tunneled into that frame, as if it were a slide under a microscope.

In the case of a falling rock, they began by freezing the picture at the instant $t = 1$ second. They wanted to know the rock's speed at that moment, but the only information they had to work with was a graph depicting time and distance. Even so, they were nearly done.

All they had to do was focus their conceptual microscope. Speed is a measure of distance traveled in a given time. *Sixty miles per hour. Three inches per second.* To solve the problem they cared about, they began by solving an easier problem, in the hope that the solution to the easy problem would point to the solution they truly wanted.

The rock's speed at a given instant was hard to find because that speed was constantly changing. But the rock's *average* speed over any particular span of time was easy to find. (Just divide the distance the rock fell by the length of the time span.) With that in mind, Newton and Leibniz did something clever. They put the actual rock to one side for a moment and concentrated instead on an easier-to-deal-with imaginary rock. The great virtue of this imaginary rock was that, in contrast with a real rock, it fell at a constant speed. What speed to pick?

The answer, Newton and Leibniz decided, was that the imaginary rock should fall at a steady speed that exactly matched the average speed of the actual rock in the interval between $t = 1$ and $t = 2$. This roundabout procedure seems like a detour, but in fact it brought them closer to their goal.

Look at the graph below. The dotted line depicts the imaginary rock, the curve the real rock. At the one-second mark (in other words, at $t = 1$) the imaginary rock and the real one have both fallen 16 feet. At $t = 2$, both the imaginary rock and the real one have fallen 64 feet.

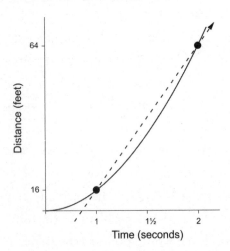

The dotted line represents the fall of an imaginary rock traveling at constant speed. The slope of the dotted line gives the imaginary rock's speed in the one-second interval between $t = 1$ and $t = 2$.

The dotted line is straight. That's crucial. Why? Because it means we can talk about its slope, which is a number—a regular, run-of-the-mill number, not an infinitesimal or any other colorful beast. That number is the speed of the imaginary rock. (It is easy to compute. Slope is a measure of steepness, which means that it is a ratio of vertical change to horizontal change. In this case, the vertical change was from 16 feet to 64 feet, and the horizontal change was from 1 second to 2 seconds, so the slope was [64–16] feet ÷ [2–1] seconds, or 48 feet per second.)

Now Newton and Leibniz made their big move. Forty-eight feet per second was the imaginary rock's speed over a one-second span. That gave a fair approximation to what they really wanted to know about, an actual rock's speed at the precise instant $t = 1$.

How could you get a better approximation? By zooming in for a closer look at the graph. And the way to do that was once again to focus your attention at $t = 1$ but this time to look at a shorter time interval than one second. As usual, pictures came to the rescue.

Look at the diagram below. The new, dashed line represents the path of a new imaginary rock. *This* imaginary rock, too, is falling at a constant speed. What speed? Not the same speed as the first imaginary rock. This new imaginary rock is falling at a speed exactly equal to the actual rock's average speed in a newer, shorter time interval, the interval between $t = 1$ and $t = 1\frac{1}{2}$. The point is that the speed of this new imaginary rock gives us a better estimate of the actual rock's speed at the instant $t = 1$.

*   *   *

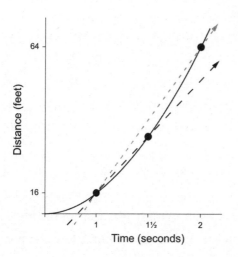

The dashed line represents the fall of a new imaginary rock. The slope of the line gives the imaginary rock's speed, which is constant, in the one-half second interval between $t = 1$ and $t = 1\frac{1}{2}$.

If we zoomed in on an even shorter interval starting at $t = 1$, we could draw still another straight line. We might, for instance, focus on the interval between $t = 1$ second and $t = 1\frac{1}{4}$ seconds. The new line, too, would have a slope that we could compute. We could repeat the procedure still another time, this time focusing on a yet-shorter interval, say between $t = 1$ second and $t = 1\frac{1}{8}$ seconds. And so on.

Newton and Leibniz saw that you could continue drawing new straight lines *forever*. Pictorially, you would be drawing straight lines that passed through two dots on the curve. One dot was fixed in place at $t = 1$, and the other moved down the curve, like a bead on a wire, approaching ever nearer to the fixed dot.

Those lines would approach ever nearer to one particular straight line. That "target" line was unique, in a natural way—it was the line that just grazed the curve at a single place, the point corresponding to $t = 1$. The target line—in mathematical jargon the tangent line—was the prize that all the fuss was about. (In the diagram below, the tangent line is the straight line made up of short dashes.) Until this moment, mathematicians had never managed to close their fingers around the notion of instantaneous speed. Now they had it.

This was an enormous breakthrough, and perhaps a recap is in order to make sure we see just what Newton and Leibniz had done. They had found a way to define a moving object's speed at a given instant. *Instantaneous speed was the number that average speeds approached, as you looked at shorter and shorter time intervals.*

Instantaneous speed wasn't a paradoxical idea or an arcane one. You could grab it and examine it at your leisure. The speed of a moving object at a given instant was just an ordinary number, the slope of the tangent line at that point. How did you

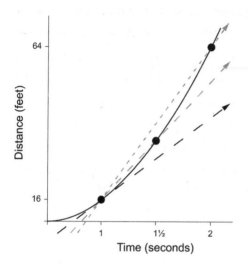

The slope of the tangent line (short dashes) represents the speed of a falling rock at the instant t = 1 second.

compute that slope? By looking at the slopes of the straight lines that approached the tangent line and seeing if those numbers approached a limit. That limit was the number we were after, the grail in this long quest.

In the case of Galileo's rock, Newton and Leibniz found that the rock's speed at the instant when it had been falling for 1 second was precisely 32 feet per second. They discovered an array of tricks that made such calculations easy whenever you had an equation to work with. Nearly always, you did. (I will skip the procedure, but it is a hint at how neatly things work out that the number they ended up with in the Galileo example—32—can be written as 16 × 2, and there were both a 16 and a 2 lurking in the equation of the curve they started with, $d = 16\,t^2$.)

Better yet, the same calculation that revealed the rock's speed at a single instant also told its speed at *every* instant. Without bothering to lift a finger or draw another straight line (let alone an infinite sequence of straight lines homing in on a target line), this once-and-for-all calculation showed that the rock's speed at any time *t* was precisely 32*t*. The speed was always changing, but a single formula captured all the changes. When the rock had been falling for 2 seconds, its speed was 64 feet per second (32 × 2). At 2½ seconds, its speed was 80 feet per second (32 × 2½); at three seconds, 96 feet per second, and so on.

This new tool for describing the moving, changing world was called calculus. With its discovery, every scientist in the world suddenly held in his hands a magical machine. Pose a question that asked *how far? how fast? how high?* and then press a button, and the machine spit out the answer. Calculus made it easy to take a snapshot—to freeze the action at any given instant—and then to examine, at leisure, an arrow momentarily motionless against the sky or an athlete hovering in midleap.

Questions that had been out of reach forever now took only a moment. How fast is a high diver traveling when she hits the water? If you shoot a rifle with the barrel at a given angle, how far will the bullet travel? What will its speed be when it reaches its target? If a drunken reveler shoots a pistol in the air to celebrate, how high will the bullet rise? More to the point, how fast will it be traveling when it returns to the ground?

Calculus was "the philosopher's stone that changed everything it touched to gold," one historian wrote, and he seemed almost resentful of the new tool's power. "Difficulties that would have baffled Archimedes were easily overcome by men not worthy to strew the sand in which he traced his diagrams."

WHEN THE CABLE SNAPS

If infinity had not always inspired terror, like some mythological dragon blocking access to a castle, someone would have discovered calculus long before Newton and Leibniz. They did not slay the dragon—the crucial concepts in calculus all hinge on infinity—but they did manage to capture and tame it. Their successors harnessed it to a plow and set it to work. For the next two centuries, science would consist largely of finding ways to exploit the new power that calculus provided. Patterns once invisible to the naked eye now showed up in vivid color. Galileo had expended huge amounts of effort to come up with the law of falling bodies, for example, but his $d = 16\ t^2$ equation contained far more information than he ever knew. Without calculus he could not see it. With calculus, there was no missing it.

Galileo knew that his law described position; he didn't know that it contained within itself a hidden law that described speed. Better yet, the law describing position was complicated, the law describing speed far simpler. In words, Galileo's position law says that after t seconds have passed, an object's distance from its starting point is $16\ t^2$ feet. It is the t^2 in that equation, rather than a simple t, that makes life complicated. As we have seen, calculus takes that law and, with only the briefest of calculations, extracts from it a new law, this one for the speed of a falling object. In words, when an object has been falling for t seconds, its speed is exactly $32t$ feet per second. In symbols (using v for velocity), $v = 32t$.

That tidy speed equation contains three surprises. First, it's simple. There's no longer any need to worry about messy numbers like t^2. Plain old t will do. Second, it holds for *every* falling object, pebbles and meteorites alike. Third, this single equation tells you a falling object's speed at *every* instant t, whether t represents 1 second or 5.3 seconds or 50. There's never any need to switch to a new equation or to modify this one. For a complete description of falling objects, this is the only equation you'll ever need.

We began with a law that described the position of a falling body and saw that it concealed within itself a simpler law describing speed. Scientists looked at that speed law and saw that it, too, concealed within itself a simpler law. And *that* one, that gem inside a gem, is truly a fundamental insight into the way the world works.

What is speed? It's a measure of how fast you're changing position. It is, to put it in a slightly more general way, a rate of change. (To be barreling down the highway at 80 miles per hour means that you're changing position at a rate of 80 miles every hour.) If we repeat the same process, by starting with speed and looking at *its* rate of change—in other words, if we compute the falling rock's acceleration—what do we find?

We find good news. Calculus tells us, literally at a glance, that a falling rock's acceleration never changes. Unlike position, that is, which depends on time in a complicated way, and unlike speed, which depends on time in a simpler way, acceleration doesn't depend on time at all. Whether a rock has been falling for one second or ten, its acceleration is always the same. It is always 32 feet per second per second. Every second that a rock continues to fall, in other words, its speed increases by another 32 feet per second. *This* is nature's doubly concealed secret.

Time (in seconds)	Position (in feet)	Speed (in feet per second)	Acceleration (in feet per second per second)
1	16	32	32
2	64	64	32
3	144	96	32

When a rock falls, its position changes in a complicated way, its velocity in a simpler way, and its acceleration in the simplest possible way.

There is a pattern in the position column, but it hardly blazes forth. The pattern in the speed column is less obscure. The pattern in the acceleration column is transparent. What do all falling objects have in common? Not their weight or color or size. Not the height they fall from or the time they take to reach the ground or their speed on impact or their greatest speed. What is true of all falling objects—an elevator snapping its cables, an egg slipping through a cook's fingers, Icarus with the wax melting from his wings—is that they all accelerate at precisely the same rate.

Acceleration is a familiar word ("the acceleration in my old car was just pitiful"), but it is a remarkably abstract notion. "It is not a fundamental quantity, such as length or mass," writes the mathematician Ian Stewart. "It is a rate of change. In fact, it is a 'second order' rate of change—that is, a rate of change of a rate of change."

Acceleration is a measure of how fast velocity is changing, in other words, and that's tricky, because velocity is a measure of how fast position is changing. "You can work out distances with a tape measure," Stewart goes on, "but it is far harder to work out a rate of change of a rate of change of distance. This is why it took humanity a long time, and the genius of a Newton, to discover the law of motion. If the pattern had been an obvious feature of distances, we would have pinned motion down a lot earlier in our history."

Acceleration turns out to be a fundamental feature of the world—unless we understand it, whole areas are off-limits to us—but it does not correspond to anything tangible. We can run a finger across a pineapple's prickly surface or heft a brick or feel the heat of a cup of coffee even through gloved hands. We could put the brick on a scale or take a ruler and measure it. Acceleration seems different from such things as the pineapple's texture and the brick's weight. We *can* measure it, but only in an indirect and cumbersome way, and we cannot quite touch it.

But it is this elusive, abstract property, Newton and Leibniz showed, that tells us how objects fall. Once again, seeing nature's secrets required looking through a mathematical lens.

Calculus had still more riches to offer. It not only revealed that distance, speed, and acceleration were all closely linked, for instance, but also showed how to move from any one of them to any of the others. That was important practically—if you wanted to know about speed, say, but you only had tools to measure time and distance, you could still find all the information you wanted, and you could do it easily—and it was important conceptually. Galileo invested endless hours in showing that if you shoot an arrow or throw a ball it travels in a parabola. Newton and Leibniz reached the same conclusion with hardly any work. All they had to know was that falling objects accelerate at 32 feet per second per second. That single number, decoded with calculus's aid, tells you almost at once that cannonballs and arrows and leaping kangaroos all travel in parabolas.

Again and again, simple observations or commonplace equations transformed themselves into wondrous insights, the mathematical counterpart of Proust's "little pieces of paper" that "the moment they are immersed in [water] stretch and shape themselves, color and differentiate, become flowers, houses, human figures, firm and recognizable."

Calculus was a device for analyzing how things change as time passes. Just what those things were made no difference. How long will it take the world's population to double? How many thousands of years ago was this mummy sealed in his tomb? How soon will the oyster harvest in the Chesapeake Bay fall to zero?

Questions about bests and worsts, when this quantity was at a maximum or that one at a minimum, could also be readily answered. Of all the roller-coaster tracks that start at a peak here and end at a valley there, which is fastest? Of all the ways to fire a cannon at a fortress high above it on a mountain, which would inflict the heaviest damage? (This was Halley's contribution, written almost as soon as he had heard of calculus. It turns out that he had also found the best angle to shoot a basketball so that it swishes through the hoop.) Of all the shapes of soap bubble one can imagine, which encloses the greatest volume with the least surface? (Nature chooses the ideal solution, a spherical bubble.) Of all the ticket prices a theater could charge, which would bring in the most money?

Not every situation could be analyzed using the techniques

of calculus. If in a tiny stretch of time a picture changed only a tiny bit, then calculus worked perfectly. From one millisecond to the next, for instance, a rocket or a sprinter advanced only a tiny distance, and calculus could tell you everything about their paths. But in the strange circumstances where something shifts abruptly, where the world jumps from one state to a different one entirely without passing through any stages in between, then calculus is helpless. (If you're counting the change in your pocket, for example, no coin is smaller than a penny, and so you jump from "twelve cents" to "thirteen cents" to "fourteen cents," with nothing in between.) One of the startling discoveries of twentieth-century science was that the subatomic world works in this herky-jerky way. Electrons jump from here to there, for instance, and in between they are . . . nowhere. Calculus throws up its hands.

But in the world we can see, most change is smooth and con-tinuous. And whenever anything changes smoothly—when a boat cuts through the water or a bullet slices through the air or a comet speeds across the heavens, when electricity flows or a cup of coffee cools or a river meanders or the high, quavering notes of a violin waft across a room—calculus provides the tools to probe that change.

Scientists wielding the new techniques talked as if they had witnessed sorcery. The old methods compared to the new, one dazed astronomer exclaimed, "as dawn compares to the bright light of noon."

Chapter Forty-Three

THE BEST OF ALL POSSIBLE FEUDS

For a long while, Newton and Leibniz spoke of one another in the most flattering terms. Newton wrote Leibniz a friendly letter in 1693, nearly a decade after Leibniz had claimed calculus for himself, hailing Leibniz as "one of the chief geometers of this century, as I have made known on every occasion that presented itself." Surely, Newton went on, there was no need for the two men to squabble. "I value my friends more than mathematical discoveries," the friendless genius declared.

Leibniz was even more effusive. In 1701, at a dinner at the royal palace in Berlin, the queen of Prussia asked Leibniz what Newton had achieved. "Taking Mathematicks from the beginning of the world to the time of Sir Isaac," Leibniz replied, "what he had done was much the better half."

But the kind words were a sham. For years, both rivals had carefully praised one another on the record while slandering each other behind the scenes. Each man composed detailed, malicious attacks on the other and published them anonymously. Each whispered insults and accusations into the ears of colleagues and then professed shock and dismay at hearing his own words parroted back.

The two geniuses had admired one another, more or less, until they realized they were rivals. Newton had long thought of the multitalented Leibniz as a dabbler in mathematics, a brilliant beginner whose genuine interests lay in philosophy and law. Leibniz had no doubts about Newton's mathematical prowess, but he believed that Newton had focused his attention in one specific, limited area. That left Leibniz free to pursue calculus on his own, or so he believed.

By the early 1700s, the clash had erupted into the open. For the next decade and a half, the fighting would grow ever fiercer. Two of the greatest thinkers of the age both clutched the same golden trophy and shouted, "Mine!" Both men were furious, indignant, unrelenting. Each felt sure the other had committed theft and compounded it with slander. Each was convinced his enemy had no motive beyond a blind lust for acclaim.

Because calculus was the ideal tool to study the natural world, the debate spilled over from mathematics to science and then from science to theology. What was the nature of the universe? What was the nature of God, who had designed that universe? Almost no one could understand the technical issues, but everyone enjoyed the sight of intellectual titans grappling like mud wrestlers. Coffeehouse philosophers weighed in; dinner parties bubbled over with gossip and delicious rumor; aristocrats across Europe chortled over the nastiest insults; in England even the royal family grew deeply involved, reviewing tactics and egging on the combatants. What began as a philosophers' quarrel grew and transmogrified until it became, in the words of the historian Daniel Boorstin, "the spectacle of the century."

*   *   *

Royalty came into the story—and threw an even brighter spotlight on Newton and Leibniz—because of Europe's complicated dynastic politics. When England's Queen Anne died without an heir, in 1714, the throne passed not to Anne's nearest relative but, so great was the fear of Catholic power, to her nearest *Protestant* relative. This was a fifty-four-year-old German nobleman named Georg Ludwig, Duke of Hanover, a brave, bug-eyed ex-soldier of no particular distinction. In England Georg Ludwig would rule as King George I.

Fond of women and cards but little else, the future king had, according to his mother, "round his brains such a thick crust that I defy any man or woman ever to discover what is in them." No matter, for Georg Ludwig had the next best thing to brains of his own. He had Europe's most renowned intellectual, Gottfried Wilhelm Leibniz, permanently on tap and at the ready.

For nearly forty years, Leibniz had served Georg Ludwig (and his father before him and that father's brother before *him*), as historian, adviser, and librarian in charge of cataloging and enlarging the ducal book collection. Among his other tasks, Leibniz had labored to establish the Hanoverian claim to the English throne. Now, with his patron suddenly plucked from the backwaters of Germany and dropped into one of the world's plum jobs, Leibniz saw a chance to return to a world capital. He had visions of accompanying his longtime employer, taking his proper place on a brightly lit stage, and trading ideas with England's greatest thinkers. Georg Ludwig had a different vision.

By the time of King George's coronation, Isaac Newton had long since made his own dazzling ascent. In 1704, he had published his second great work, *Opticks*, on the properties of light. In 1705, the onetime farmboy had become Sir Isaac Newton, the first scientist ever knighted. (Queen Anne had performed the ceremony. Anne was no scholar—"When in good humour Queen Anne was meekly stupid, and when in bad humor, was sulkily stupid," the historian Macaulay had observed—but she had savvy counselors who saw political benefit in honoring England's greatest thinker.)

By the time of his knighthood, Newton was sixty-two and had largely abandoned scientific research. A few years before, he had left Cambridge in favor of London and accepted a government post as warden of the Mint. At roughly the same time he took on the presidency of the Royal Society, a position he would hold until his death. Old, imposing, intimidating, Newton was universally hailed as the embodiment of genius. English genius, in particular. Many who could not tell a parrot from a parabola gloried in the homage paid to England's greatest son. When dignitaries like Russia's Peter the Great visited London, they made a point of seeing Newton along with the capital's other marvels.

Newton did not become much of a partygoer in his London days, but his new circle of acquaintances did come to include such ornaments as Caroline, Princess of Wales. King George himself kept a close watch on the Newton-Leibniz affair. His motive was not intellectual curiosity—the king's only cultural interests were listening to opera and cutting out paper dolls—but he took malicious delight in having a claim on two of the greatest men of the age. King George seemed an unlikely candidate to preside over a philosophical debate. In Germany his court had been caught up not only in scandal but quite likely in murder.

The problems rose out of a tangled series of romantic liaisons. All the important men at the Hanover court had mistresses, often several at a time, and a diagram of whose bed partner was whose would involve multiple arrows crossing one another and looping back and forth. (Adding to the confusion, nearly all the female participants in the drama seemed to share the name Sophia or some near variation.) Bed-hopping on the part of the Hanover princes fell well within the bounds of royal privilege. What was *not* acceptable was that Georg Ludwig's wife, Sophia Dorothea, had embarked on an affair of her own. Royal spies discovered that the lovers had made plans to run off together. This was unthinkable. A team of hired assassins ambushed the duchess's paramour, stabbed him with a sword, sliced him open with an axe, and left him to bleed to death. Sophia Dorothea was banished to a family castle and forbidden ever to see her children again. She died thirty-two years later, still under house arrest.

Through the years Leibniz's attempts to engage Georg Ludwig had met with about the success one would expect, but the women of the Hanoverian court were as intellectual as the men were crude. While the dukes collected mistresses and plotted murder, their duchesses occupied themselves with philosophy. Georg Ludwig's mother, Sophia, read through Spinoza's controversial writings as soon as they were published and spent long hours questioning Leibniz about the views of the Dutch heretic.

Sophia was only the first of Leibniz's royal devotees. Sophia's daughter Sophia Charlotte (sister to the future King George) had an even closer relationship with Leibniz. And yet a third high-born woman forged a still closer bond. This was Caroline, a twenty-one-year-old princess and friend of Sophia Charlotte. Leibniz became her friend and tutor. Soon after, Caroline married one of Georg Ludwig's brothers. When she was whisked off to England in 1714, Caroline became princess of Wales and in time, as the wife of King George II, queen of England. Leibniz had allies in the highest of circles.

But he was stuck in Germany, and none of his royal friends seemed inclined to send for him. From that outpost, he tried to enlist Caroline on his side in his ongoing war against Newton. Their battle represented not just a confrontation between two men, Leibniz insisted, but between two nations. German pride was at stake. "I dare say," Leibniz wrote to Caroline, "that if the king were at least to make me the equal of Mr. Newton in all things and in all respects, then in these circumstances it would give honor to Hanover and to Germany in my name."

The appeal to national pride proved ineffective. Newton was all but worshipped in England—as we have noted, Caroline had met him on various grand occasions at court—and the newly arrived king had no desire to challenge English self-regard just to soothe the hurt feelings of his pet philosopher. In any case, King George had his own plans for Leibniz. They did not include science. Leibniz's chief duty, the king reminded him, was to continue his history of the House of Hanover. He had bogged down somewhere around the year 1000.

The wonders of calculus, and the injustice of Newton's theft of it, concerned the king not at all. What was life and death for Leibniz was sport for King George. "The king has joked more than once about my dispute with Mr. Newton," Leibniz lamented.

From his exile in Hanover, Leibniz wrote to Caroline attacking Newton's views on science and theology. Caroline studied the letters intently—they dealt mainly with such questions as whether God had left the world to run on its own or whether He continued to step in to fine-tune it—and she passed them along to a Newton stand-in named Samuel Clarke. On some questions Caroline wrote directly to Newton himself. Clarke composed responses to Leibniz (with Newton's help). The correspondence was soon published, and the so-called Leibniz-Clarke papers became, in one historian's judgment, "perhaps the most famous and influential of all philosophical correspondences."

But to Caroline's exasperation, Leibniz persisted in setting aside deep issues in theology and circling back instead to his priority battle with Newton. The princess scolded her ex-tutor for his "vanity." He and Newton were "the great men of our century," Caroline wrote, "and both of you serve a king who merits you." Why draw out this endless fight? "What difference does it make whether you or Chevalier Newton discovered the calculus?" Caroline demanded.

A good question. The world had the benefit of this splendid new tool, after all, whoever had found it. But to Newton and Leibniz, the answer to Caroline's question was simple. It made all the difference in the world.

Chapter Forty-Four

BATTLE'S END

From its earliest days, science has been a dueling ground. Disputes are guaranteed, because good ideas are "in the air," not dreamed up out of nowhere. Nearly every breakthrough—the telescope, calculus, the theory of evolution, the telephone, the double helix—has multiple parents, all with serious claims. But ownership is all, and scientists turn purple with rage at the thought that someone has won praise for stolen insights. The greats battle as fiercely as the mediocre. Galileo wrote furiously of rivals who claimed that they, not he, had been first to see sunspots. They had, he fumed, "attempted to rob me of that glory which was mine." Even the peaceable Darwin admitted, in a letter to a colleague urging him to write up his work on evolution before he was scooped, that "I certainly should be vexed if anyone were to publish my doctrines before me."

What vexed the mild Darwin sent Newton and Leibniz into apoplectic rages. The reasons had partly to do with mathematics itself. All scientific feuds tend toward the nasty; feuds between mathematicians drip with extra venom. Higher mathematics is a peculiarly frustrating field. So difficult is it that even the best mathematicians often feel that the challenge is just too much, as if a golden retriever had taken on the task of understanding the workings of the internal combustion engine. The rationalizations so helpful elsewhere in science—she had a bigger lab, a larger budget, better colleagues—are no use here. Wealth, connections, charm make no difference. Brainpower is all.

"Almost no one is capable of doing significant mathematics," the American mathematician Alfred W. Adler wrote a few decades ago. "There are no acceptably good mathematicians. Each generation has its few great mathematicians, and mathematics would not even notice the absence of the others. They are useful as teachers, and their research harms no one, but it is of no importance at all. A mathematician is great or he is nothing."

That is a romantic view and probably overstated, but mathematicians take a perverse pride in great-man theories, and they tend to see such doctrines as simple facts. The result is that mathematicians' egos are both strong and brittle, like ceramics. Where they focus their gaze makes all the difference. If someone compares himself with his neighbors, then he might preen himself on his membership in an arcane priesthood. But if he judges himself not by whether he knows more mathematics than most people but by whether he has made any real headway at exploring the immense and dark mathematical woods, then all thoughts of vanity flee, and only puniness remains.

In the case of calculus, the moment of confrontation between Newton and Leibniz was delayed for a time, essentially by incredulity. Neither genius could quite believe that anyone else could have seen as far as he had. Newton enjoyed his discoveries all the more because they were his to savor in solitude, as if he were a reclusive art collector free to commune with his masterpieces behind closed doors. But Newton's retreat from the world was not complete. He could abide adulation but not confrontation, and he had shared some of his mathematical triumphs with a tiny number of appreciative insiders. He ignored their pleas that he tell everyone what he had told them. The notion that his discoveries would speed the advance of science, if only the world knew of them, moved Newton not at all.

For Leibniz, on the other hand, his discoveries had value precisely because they put his merits on display. He never tired of gulping down compliments, but his eagerness for praise had a practical side, too. Each new achievement served as a golden entry on the résumé that Leibniz was perpetually thrusting before would-be patrons.

In Newton's view, to unveil a discovery meant to offer the unworthy a chance to paw at it. In Leibniz's view, to proclaim a discovery meant to offer the world a chance to shout its hurrahs.

In history's long view, the battle ended in a stalemate. Historians of mathematics have scoured the private papers of both men and found clear evidence that Newton and Leibniz discovered calculus independently, each man working on his own. Newton was first, in 1666, but he only published decades later, in 1704. Leibniz's discovery followed Newton's by nine years, but he published his findings first, in 1684. And Leibniz, who had a gift for devising useful notations, wrote up his discoveries in a way that other mathematicians found easy to understand and build upon. (Finding the right notation to convey a new concept sounds insignificant, like choosing the right typeface for a book, but in mathematics the choice of symbols can save an idea or doom it. A child can multiply 17 by 19. The greatest scholars in Rome would have struggled with XVII times XIX.)*

---

* Sometimes the right notation can even hint at a deep, surprising insight. Simply using decimal notation, and then adding column by column, suggests that $1 + .1 + .01 + .001 + \ldots = 1.11111\ldots$, and not infinity.

The symbols and language that Leibniz devised are still the ones that students learn today. Newton's discovery was identical, at its heart, and in his masterly hands it could be turned to nearly any task. But Newton's calculus is a museum piece today, while a buffed and honed version of Leibniz's remains in universal use. Newton insisted that because he had found calculus before anyone else, there was nothing to debate. Leibniz countered that by casting his ideas in a form that others could follow, and then by telling the world what he had found, he had thrown open a door to a new intellectual kingdom.

So he had, and throughout the 1700s and into the 1800s, European mathematicians inspired by Leibniz ran far in front of their English counterparts. But in their lifetimes, Newton seemed to have won the victory. To stand up to Newton at his peak of fame was nearly hopeless. The awe that Alexander Pope would later encapsulate—"Nature and nature's laws lay hid in night, / God said 'Let Newton be!' and all was light"—had already become common wisdom.

The battle between the two men smoldered for years before it burst into open flames. In 1711, after about a decade of mutual abuse, Leibniz made a crucial tactical blunder. He sent the Royal Society a letter—both he and Newton were members—complaining of the insults he had endured and asking the Society to sort out the calculus quarrel once and for all. "I throw myself on your sense of justice," he wrote.

He should have chosen a different target. Newton, who was president of the Royal Society, appointed an investigatory committee "numerous and skilful and composed of Gentlemen of several Nations." In fact, the committee was a rubber stamp for Newton himself, who carried out its inquiry single-handedly and then issued his findings in the committee's name. The report came down decisively in Newton's favor. With the Royal Society's imprimatur, the long, damning report was distributed to men of learning across Europe. "We take the Proper Question to be not who Invented this or that Method but who was the first Inventor," Newton declared, for the committee.

The report went further. Years before, it charged, Leibniz had been offered surreptitious peeks at Newton's mathematical papers. There calculus was "Sufficiently Described" to enable "any Intelligent Person" to grasp its secrets. Leibniz had not only lagged years behind Newton in finding calculus, in other words, but he was a sneak and a plagiarist as well.

Next the *Philosophical Transactions*, the Royal Society's scientific journal, ran a long article reviewing the committee report and repeating its anti-Leibniz charges. The article was unsigned, but Newton was the author. Page after page spelled out the ways in which "Mr. Leibniz" had taken advantage of "Mr. Newton." Naturally Mr. Leibniz had his own version of events, but the anonymous author would have none of it. "Mr. Leibniz cannot be a witness in his own Cause."

Finally the committee report was republished in a new edition accompanied by Newton's anonymous review. The book carried an anonymous preface, "To the Reader." It, too, was written by Newton.

Near the end of his life Newton reminisced to a friend about his long-running feud. "He had," he remarked contentedly, "broke Leibniz' heart."

Chapter Forty-Five

THE APPLE AND
THE MOON

The greatest scientific triumph of the seventeenth century, Newton's theory of universal gravitation, was in a sense a vehicle for showing off the power and range of the mathematical techniques that Newton and Leibniz had fought to claim. Both men discovered calculus, but it was Newton who provided a stunning demonstration of what it could do.

Until 1687, Isaac Newton had been known mainly, to those who knew him at all, as a brilliant mathematician who worked in self-imposed isolation. No recluse ever broke his silence more audaciously.

Fame came with the publication of the *Principia*. Newton had been at Cambridge for two decades. University rules required that he teach a class or two, but this did not impose much of a burden, either on Newton or anyone else. "So few went to hear Him, & fewer that understood him," one contemporary noted, "that oftimes he did in a manner, for want of Hearers, read to ye Walls."

As Newton told the story, his rise to fame had indeed begun with the fall of an apple. In his old age he occasionally looked back on his career, and eager listeners noted down every word. A worshipful young man named John Conduitt, the husband of Newton's niece, was one of several who heard the apple story firsthand. "In the year 1666 he retired again from Cambridge . . . to his mother in Lincolnshire," Conduitt wrote, "& whilst he was musing in a garden it came into his thought that the power of gravity (which brought an apple from the tree to the ground) was not limited to a certain distance from the earth but that this power must extend much farther than was usually thought. Why not as high as the moon said he to himself & if so that must influence her motion & perhaps retain her in her orbit, whereupon he fell a calculating. . . ."

The story, which is the one thing everyone knows about Isaac Newton, may well be a myth.* Despite his craving for privacy, Newton was acutely aware of his own legend, and he was not above adding a bit of gloss here and there. Historians who have scrutinized his private papers believe that his understanding of gravity dawned slowly, over several years, rather than in a flash of insight. He threw in the apple, some suspect, simply for color.

In any case, it wouldn't have taken an apple to remind Newton that objects fall. Everyone had always known that. The point was to look beyond that fact to the questions it raised. If apples fell to the ground because some force drew them, did that force extend from the tree's branches to its top? And beyond the top to . . . to where? To the top of a mountain? To the clouds? To the moon? Those questions had seldom been asked. There were many more. What about the apple when it was *not* falling? An apple in a tree stays put because it is attached to the branch. No

---

* Most people "know" not just that an apple fell but that it bonked Newton on the head.

surprise there. But what about the moon? What holds the moon up in the sky?

Before Newton the answer had two parts. The moon stayed in the sky because that was its natural home and because it was made of an ethereal substance that was nothing like the heavy stuffing of bodies here on Earth. But that would no longer do. If the moon was just a big rock, as telescopes seemed to show, why didn't it fall like other rocks?

The answer, Newton came to see, was that *it does fall*. The breakthrough was to see how that could be. How could something fall and fall but never arrive? Newton's answer in the case of the moon, a natural satellite, ran much like the argument we have already seen, for an artificial satellite.

We tend to forget the audacity of that explanation, and Newton's plain tone helps us along in our mistake. "I began to think of gravity extending to ye orb of the Moon," he recalled, as if nothing could have been more natural. Newton began to give serious thought, in other words, to asking whether the same force that pulled an apple to the Earth also pulled the moon toward the Earth. But this is to downplay two feats of intellectual daring. Why should anyone have thought that the moon is falling, first of all, when it is plainly hanging placidly in the sky, far beyond our reach or the reach of anything else? And even if we did make the large concession that it is falling, second of all, why should that fall have anything in common with an apple's fall? Why would anyone presume that the same rules governed realms as different as heaven and Earth?

But that is exactly what Newton did presume, for aesthetic and philosophical reasons as much as for scientific ones. Throughout his life Newton believed that God operated in the simplest, neatest, most efficient way imaginable. That principle served as his starting point whether he was studying the Bible or the natural world. (We have already noted his insistence that "it is ye perfection of God's works that they are all done with ye greatest simplicity.") The universe had no superfluous parts or forces for exactly the reason that a clock had no superfluous wheels or springs. And so, when Newton's thoughts turned to gravity, it was all but inevitable that he would wonder how much that single force could explain.

Newton's first task was to find a way to turn his intuition about the sweep and simplicity of nature's laws into a specific, testable prediction. Gravity certainly seemed to operate here on Earth; if it did reach all the way to the moon, how would you know it? How would gravity reveal itself? To start with, it seemed clear that if gravity did extend to the moon, its force must diminish over that vast distance. But how much? Newton had two paths to an answer. Fortunately, both gave the same result.

First, he could try intuition and analogy. If we see a bright light ten yards off, say, how bright will it be if we move it twice as far away, to twenty yards distance? The answer was well-known. Move a light twice as far away and it will not be half as bright, as you might guess, but only one-fourth as bright. Move it ten times as far away and it will be one-hundredth as bright. (The reason has to do with the way light spreads. Sound works the same way. A piano twenty yards away sounds only one-fourth as loud as a piano ten yards away.)

So Newton might have been tempted to guess that the pull of gravity decreases with distance in the same way that the brightness of light does. Physicists today talk about "inverse-square laws," by which they mean that some forces weaken not just in proportion to distance but in proportion to distance squared.

(It would later turn out that electricity and magnetism follow inverse-square laws, too.)

A second way of looking at gravity's pull gave the same answer. By combining Kepler's third law, which had to do with the size and speed of the planets' orbits, with an observation of his own about objects traveling in a circle, Newton calculated the strength of gravity's pull. Again, he found that gravity obeyed an inverse-square law.

Now came the test. If gravity actually pulled on the moon, how much did it pull? Newton set to work. He knew that the moon orbits the Earth. It travels in a circle, in other words, and not in a straight line. (To be strictly accurate, it travels in an ellipse that is almost but not quite circular, but the distinction does not come into play here.) He knew, as well, what generations of students have since had drummed into them as "Newton's first law"—in modern terms, *a body in motion will travel in a straight line at a steady speed unless some force acts on it (and a body at rest will stay at rest unless some force acts on it).*

So some force was acting on the moon, pulling it off a straight-line course. How far off course? That was easy to calculate. To start with, Newton knew the size of the moon's orbit, and he knew that the moon took a month to travel once around that circuit. Taken together, those facts told him the moon's speed. Next came a thought experiment. What would happen to the moon if gravity were magically turned off for a second? Newton's first law gave him the answer—it would shoot off into space on a straight line, literally going off on a tangent. (If you tied a rock with a piece of string and swung it around your head, the rock would travel in a circle until the string snapped, and then it would fly off in a straight line.)

But the moon stays in its circular orbit. Newton knew what

that meant. It meant a force was pulling it. Now he needed some numbers. To find out how far the moon was pulled, all he had to do was calculate the distance between where the moon actually is and where it would have been if it had traveled in a straight line. That distance was the fall Newton was looking for—the moon "falls" from a hypothetical straight line to its actual position.

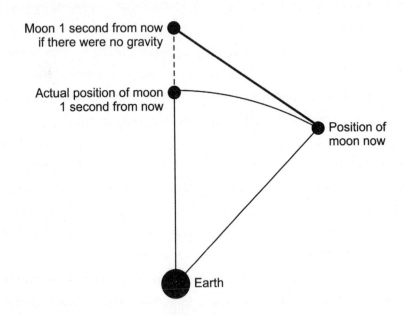

Newton calculated the distance the moon falls in 1 second,
which corresponds to the dashed line in the diagram.

In his quest to compare Earth's pull on the moon and on an apple, Newton was nearly home. He knew how far the moon falls in one second. He had just calculated that. It falls about $\frac{1}{20}$ of an inch. He knew how far an apple falls in one second. Galileo had found that out, with his ramps: 16 feet.

All that remained was to look at the ratio of those two falls, the ratio of $\frac{1}{20}$ of an inch to 16 feet. The last puzzle piece was the distance from the Earth to the moon. Why did that matter?

Because the distance from the Earth to the moon was about 60 times the distance from the center of the Earth to the Earth's surface. Which was to say that the moon was 60 times as far from the center of the Earth as the apple was. If gravity truly did follow an inverse-square law, then the Earth's pull on the moon should be 3,600 times weaker (60 × 60) than its pull on the apple.

Only the last, crucial calculation remained. The moon fell $\frac{1}{20}$ of an inch in one second; an apple fell 16 feet in one second. Was the ratio of $\frac{1}{20}$ of an inch to 16 feet the same as the ratio of 1 to 3,600, as Newton had predicted? How did the moon's fall compare with the apple's fall?

Just as Newton had hoped it would, or nearly so. The two ratios almost matched. Newton "compared the force required to keep the Moon in her Orb with the force of gravity," he wrote proudly, "& found them answer pretty nearly." The same calculation carried out today, with far better data than Newton had available, would give even closer agreement. That wasn't necessary. The big message was already clear. Gravity reached from the Earth to the moon. The same force that drew an apple drew the moon. The same law held here and in the heavens. God had indeed designed his cosmos with "ye greatest simplicity."

Chapter Forty-Six

A VISIT TO CAMBRIDGE

Newton's moon calculation had buttressed his faith in simple laws, but he still had an immense distance to cover before he could prove his case. The moon was not the universe. What of Kepler's laws, for instance? The great astronomer had devoted his life to proving that the planets traveled around the sun in ellipses. How did ellipses fit in God's cosmic architecture?

Stymied by the difficulty of sorting out gravity, or perhaps tempted more by questions in other fields, Newton had put gravity aside after his miracle years. He had made his apple-and-moon calculation when he was in his twenties. For the next twenty years he gave most of his attention to optics, alchemy, and theology instead.

Late on a January afternoon in 1684, Robert Hooke, Christopher Wren, and Edmond Halley left a meeting of the Royal Society and wandered into a coffeehouse to pick up a conversation they had been carrying on all day. Coffee had reached England only a generation before, but coffeehouses had spread everywhere.* Hooke in particular seemed to thrive in the rowdy

---

* Another exotic import, tea, had arrived at about the same time, although coffee caught on first. On September 25, 1660, Pepys wrote in his diary that "I did send for a cup of tee (a China drink) of which I never had drank before."

atmosphere. In crowded rooms thick with the hubbub of voices and the smells of coffee, chocolate, and tobacco, men sat for hours debating business, politics, and, lately, science. (Rumors and "false news" spread so quickly, as with the Internet today, that the king tried, unsuccessfully, to shut coffeehouses down.)

With steaming mugs in hand, the three men resumed talking of astronomy. All three had already guessed, or convinced themselves by the same argument Newton had made using Kepler's third law, that gravity obeyed an inverse-square law. Now they wanted the answer to a related question—if the planets did follow an inverse-square law, what did that tell you about their orbits? This question—in effect, *where do Kepler's laws come from?*—was one of the central riddles confronting all the era's scientists.

Halley, a skilled mathematician, admitted to his companions that he had tried to find an answer and failed. Wren, still more skilled, confessed that *his* failures had stretched over the course of several years. Hooke, who was sometimes derided as the "universal claimant" for his habit of insisting that every new idea that came along had occurred to him long before, said that he'd solved this problem, too. For the time being, he said coyly, he preferred to keep the answer to himself. "Mr. Hook said that he had it," Halley recalled later, "but that he would conceale it for some time, that others trying and failing might know how to value it, when he should make it publick."

Wren, dubious, offered a forty-shilling prize—roughly four hundred dollars today—to anyone who could find an answer within two months. No one did. In August 1684, Halley took the question to Isaac Newton. Halley, one of the few great men of the Royal Society who was charming as well as brilliant, scarcely knew Newton, though he knew his mathematical reputation. But Halley could get along with everyone, and he made a perfect ambassador. Though still only twenty-eight, he had already made his mark in mathematics and astronomy. Just as important, he was game for anything. In years to come he would stumble through London's taverns with Peter the Great, on the czar's visit to London; he would invent a diving bell (in the hope of salvaging treasure from shipwrecks) and would descend deep underwater to test it himself; he would tramp up and down mountains to compare the barometric pressure at the summit and the base; in an era of wooden ships he would survey vast swaths of the world's oceans, from the tropics to "islands of ice."

Now his task was to win over Isaac Newton. "After they had been some time together," as Newton later told the story to a colleague, Halley explained the reason for his visit. He needed Newton's help. The young astronomer spelled out the question that had stumped him, Wren, and Hooke. If the sun attracted the planets with a force that obeyed an inverse-square law, what shape would the planets' orbits be?

"Sir Isaac replied immediately that it would be an Ellipsis." Halley was astonished. "The Doctor struck with joy & amazement asked him how he knew it. Why saith he I have calculated it."

Halley asked if he could see the calculation. Newton rummaged through his papers. Lost. Halley extracted a promise from Newton to work through the mathematics again, and to send him the results.

Chapter Forty-Seven

NEWTON
BEARS DOWN

The paper was not really lost. Newton, the most cautious of men, wanted to reexamine his work before he revealed it to anyone. Looking over his calculations after Halley's visit, Newton did indeed catch a mistake. He corrected it, expanded his notes, and, three months later, sent Halley a formal, nine-page treatise, in Latin, titled "On the Motion of Bodies in an Orbit." It did far, far more than answer Halley's question.

Kepler's discovery that the planets travel in ellipses, for instance, had never quite made sense. It was a "law" in the sense that it fit the facts, but it seemed dismayingly arbitrary. Why ellipses rather than circles or figure eights? No one knew. Kepler had agonized over the astronomical data for years. Finally, for completely mysterious reasons, ellipses had turned out to be the curves that matched the observations. Now Newton explained where ellipses came from. He showed, using calculus-based arguments, that if a planet travels in an ellipse, then the force that attracts it *must* obey an inverse-square law. The flip side was true, too. If a planet orbiting around a fixed point does obey an inverse-square law, then it travels in an ellipse.* All this was a matter of strict mathematical fact. Ellipses and inverse-square laws were intimately connected, though it took Newton's genius to see it, just as it had taken a Pythagoras to show that right triangles and certain squares were joined by hidden ties.

Newton had solved the mystery behind Kepler's second law, as well. It, too, summarized countless astronomical observations in one compact, mysterious rule—planets sweep out equal areas in equal times. In his short essay, Newton deduced the second law, as he had deduced the first. His tools were not telescope and sextant but pen and ink. All he needed was the assumption that some force draws the planets toward the sun. Starting from that bare statement (without saying anything about the shape of the planets' orbits or whether the sun's pull followed an inverse-square law), Newton demonstrated that Kepler's law had to hold. Mystery gave way to order.

Bowled over, Halley rushed back to Cambridge to talk to Newton again. The world needed to hear what he had found. Remarkably, Newton went along. First, though, he would need to improve his manuscript.

Thus began one of the most intense investigations in the history of thought. Since his early years at Cambridge, Newton had largely abandoned mathematics. Now his mathematical fever surged up again. For seventeen months Newton focused all his

---

* The statement *if a planet travels in an ellipse, then it follows an inverse-square law* is different from the statement *if a planet follows an inverse-square law, then it travels in an ellipse*. It might have been that one was true but the other was not. *If someone owns a dog, then he owns a pet* is true; *if someone owns a pet, then he owns a dog* is not. In this case it was clear to Newton (though bewilderingly obscure to others) that if one statement was true, the other had to be true as well.

powers on the question of gravity. He worked almost without let-up, with the same ferocious concentration that had marked his miracle years two decades before.

Albert Einstein kept a picture of Newton above his bed, like a teenage boy with a poster of LeBron James. Though he knew better, Einstein talked of how easily Newton made his discoveries. "Nature to him was an open book, whose letters he could read without effort." But the real mark of Newton's style was not ease but power. Newton focused his gaze on whatever problem had newly obsessed him, and then he refused to look away until he had seen to its heart.

"Now I am upon this subject," he told a colleague early in his investigation of gravity, "I would gladly know ye bottom of it before I publish my papers." The matter-of-fact tone obscures Newton's drivenness. "I never knew him take any Recreation or Pastime," recalled an assistant, "either in Riding out to take ye Air, Walking, Bowling, or any other Exercise whatever, thinking all Hours lost that was not spent in his Studyes." Newton would forget to leave his rooms for meals until he was reminded and then "would go very carelessly, with Shooes down at Heels, Stockings unty'd . . . & his Head scarcely comb'd."

Such stories were in the standard vein of anecdotes about absentminded professors, already a cliché in the 1600s,* except that in Newton's case the theme was not otherworldly dreaminess but energy and singleness of vision. Occasionally a thought would strike Newton as he paced the grounds near his rooms.

---

* Sixteen hundred years before Newton, Plutarch wrote that Archimedes grew so absorbed in his thoughts that he "would often forget his food and neglect his person" and have to be "carried by absolute violence to bathe."

(It was not quite true that he never took a walk to clear his head.) "When he has sometimes taken a Turn or two he has made a sudden Stand, turn'd himself about, run up ye stairs & like another Archimedes, with a *Eureka!*, fall to write on his Desk standing, without giving himself the Leasure to draw a Chair to sit down in."

Even for Newton the assault on gravity demanded a colossal effort. The problem was finding a way to move from the idealized world of mathematics to the messy world of reality. The diagrams in Newton's "On Motion" essay for Halley depicted points and curves, much as you might see in any geometry book. But those points represented colossal, complicated objects like the sun and the Earth, not abstract circles and triangles. Did the rules that held for textbook examples apply to objects in the real world?

Newton was exploring the notion that all objects attracted one another and that the strength of that attraction depended on their masses and the distance between them. Simple words, it seemed, but they presented gigantic difficulties. What was the distance between the apple and the Earth? For two objects separated by an enormous distance, like the Earth and the moon, the question seemed easy. In that case, it hardly mattered precisely where you began measuring. For simplicity's sake, Newton took "distance" to mean the distance between the centers of the two objects. But when it came to the question of the attraction between an apple and the Earth, what did the center of the Earth have to do with anything? An apple in a tree was thousands of miles from the Earth's center. What about all those parts of the Earth that *weren't* at the center? If everything attracted everything else, wouldn't the pulls from bits of ground near the tree have to be taken into account? How would you tally up all those millions and millions of pulls, and wouldn't they combine to overcome the pull from a faraway spot like the center of the Earth?

Mass was just as bad. The Earth certainly wasn't a point, though Newton had drawn it that way. It wasn't even a true sphere. Nor was it uniform throughout. Mountains soared here, oceans swelled there, and, deep underground, strange and unknown structures lurked. And that was just on Earth. What of the sun and the other planets, and what about all their simultaneous pulls? "To do this business right," Newton wrote Halley in the middle of his bout with the *Principia*, "is a thing of far greater difficulty than I was aware of."

But Newton did do the business right, and astonishingly quickly. In April 1686, less than two years after Halley's first visit, Newton sent Halley his completed manuscript. His nine-page essay had grown into the *Principia's* five hundred pages and two-hundred-odd theorems, propositions, and corollaries. Each argument was dense, compact, and austere, containing not a spare word or the slightest note of warning or encouragement to his hard-pressed readers. The modern-day physicist Subrahmanyan Chandrasekhar studied each theorem and proof minutely. Reading Newton so closely left him more astonished, not less. "That all these problems should have been enunciated, solved, and arranged in logical sequence in seventeen months is beyond human comprehension. It can be accepted only because it is a fact."

The *Principia* was made up of an introduction and three parts, known as Books I, II, and III. Newton began his introduction with three propositions now known as Newton's laws. These were not summaries of thousands of specific facts, like Kepler's laws, but magisterial pronouncements about the behavior of nature in general. Newton's third law, for instance, was the famous "to every action, there is an equal and opposite reaction." Book I dealt essentially with abstract mathematics, focused on topics like orbits and inverse squares. Newton discussed not the crater-speckled moon or the watery Earth but a moving point P attracted toward a fixed point S and moving in the direction AB, and so on.

In Book II Newton returned to physics and demolished the theories of those scientists, most notably Descartes, who had tried to describe a mechanism that accounted for the motions of the planets and the other heavenly bodies. Descartes pictured space as pervaded by some kind of ethereal fluid. Whirlpools within that fluid formed "vortices" that carried the planets like twigs in a stream. Something similar happened here on Earth; rocks fell because mini-whirlpools dashed them to the ground.

Some such "mechanistic" explanation had to be true, Descartes insisted, because the alternative was to believe in magic, to believe that objects could spring into motion on their own or could move under the direction of some distant object that never came in contact with them. That couldn't be. Science had banished spirits. The only way for objects to interact was by making contact with other objects. That contact could be direct, as in a collision between billiard balls, or by way of countless, intermediate collisions with the too-small-to-see particles that fill the universe. (Descartes maintained that there could be no such thing as a vacuum.)

Much of Newton's work in Book II was to show that Descartes' model was incorrect. Whirlpools would eventually fizzle out. Rather than carry a planet on its eternal rounds, any whirl-pool would sooner or later be "swallowed up and lost." In any case, no such picture could be made to fit with Kepler's laws.

Then came Book III, which was destined to make the *Principia* immortal.

Chapter Forty-Eight

TROUBLE WITH MR. HOOKE

If not for the *Principia*'s unsung hero, Edmond Halley, the world might never have seen Book III. At the time he was working to coax the *Principia* from Newton, Halley had no official standing to speak of. He was a minor official at the Royal Society—albeit a brilliant scientist—who had taken on the task of dealing with Newton because nobody else seemed to be paying attention. Despite its illustrious membership, the Royal Society periodically fell into confusion. This was such a period, with no one quite in charge and meetings often canceled.

So the task of shepherding along what would become one of the most important works in the history of science fell entirely to Halley. It was Halley who had to deal with the printers and help them navigate the impenetrable text and its countless abstruse diagrams, Halley who had to send page proofs to Newton for his approval, Halley who had to negotiate changes and corrections. Above all, it was Halley who had to keep his temperamental author content.

John Locke once observed that Newton was "a nice man to deal with"—"nice" in the seventeenth-century sense of "finicky"—

which was true but considerably understated. Anyone dealing with Newton needed the delicate touch and elaborate caution of a man trying to disarm a bomb. Until he picked up the *Principia* from the printer and delivered the first copies to Newton, Halley never dared even for a moment to relax his guard.

On May 22, 1686, after Newton had already turned in Books I and II of his manuscript, Halley worked up his nerve and sent Newton a letter with unwelcome news. "There is one thing more I ought to informe you of," he wrote, "viz, that Mr Hook has some pretensions upon the invention of ye rule of the decrease of Gravity. . . . He says you had the notion from him." Halley tried to soften the blow by emphasizing the limits of Hooke's claim. Hooke maintained that he had been the one to come up with the idea of an inverse-square law. He conceded that he had not seen the connection between inverse squares and elliptical orbits; that was Newton's insight, alone. Even so, Halley wrote, "Mr Hook seems to expect you should make some mention of him."

Instead, Newton went through the *Principia* page by page, diligently striking out Hooke's name virtually every time he found it. "He has done nothing," Newton snarled to Halley. Newton bemoaned his mistake in revealing his ideas and thereby opening himself up to attack. He should have known better. "Philosophy [i.e., science] is such an impertinently litigious Lady that a man had as good be engaged in Law suits as have to do with her," he wrote. "I found it so formerly & now I no sooner come near her again but she gives me warning."

The more Newton brooded, the angrier he grew. Crossing out Hooke's name was too weak a response. Newton told Halley that he had decided not to publish Book III. Halley raced to soothe Newton. He could not do without Newton's insights; the Royal Society could not; the learned world could not.

* * *

Newton could have dismissed the controversy with a gracious tip of the hat to Hooke, for Hooke had indeed done him a favor. In 1684, as we have seen, Halley had asked Newton a question about the inverse-square law, and Newton had immediately given him the answer.

The reason Newton knew the answer is that Hooke had written him a letter four years before that asked the identical question. What orbit would a planet follow if it were governed by an inverse-square law? "I doubt not but that by your excellent method you will easily find out what that Curve must be," Hooke had written Newton, "and its proprietys [properties], and suggest a physicall Reason of this proportion."

Newton had solved the problem then and put it away. He never replied to Hooke's letter. This was perhaps inevitable, for Hooke and Newton had been feuding for years. Back in 1671, the Royal Society had heard rumors of a new kind of telescope, supposedly invented by a young Cambridge mathematician. The rumors were true. Newton had designed a telescope that measured a mere six inches but was more powerful than a conventional telescope six feet long. The Royal Society asked to see it, Newton sent it along, and the Society oohed and aahed.

Newton's reputation was made. This was Newton's first contact with the Royal Society, which at once invited him to join. He accepted. Only Hooke, until this new development England's unchallenged authority on optics and lenses, refused to add his voice to the chorus of praise.

Even a better-natured man than Hooke might have bristled at all the attention paid to a newcomer (Hooke was seven years older than Newton), but Hooke was fully as proud and prickly as Newton himself. In 1671 Hooke was an established scientific

figure; Newton was unknown. Hooke had spent a career craft-
ing instruments like the telescopes that Newton's new design
had so dramatically surpassed; Newton's main interests were in
other areas altogether. And more trouble lay just ahead, though
Hooke could not have anticipated it. In a letter to the Royal
Society thanking them for taking such heed of his telescope,
Newton added a tantalizing sentence. In the course of his "poore
& solitary endeavours," he had found something remarkable.

Within a month, Newton followed up his coup with the telescope by sending the Royal Society his groundbreaking paper on
white light. The nature of light was another of Hooke's particular interests. Once again, the outsider had barged into staked-
out territory and put down his own marker. Deservedly proud
of what he had found, Newton for once said so openly. His demonstration that white light was made up of all the colors was,
Newton wrote, "the oddest, if not the most considerable detec-
tion, which has hitherto been made in the operation of nature."

The paper, later hailed as one of the all-time landmarks in science, met with considerable resistance at first, from Hooke most
of all. He had already done all of the same experiments, Hooke
claimed, and, unlike Newton, he had interpreted them correctly.
He said so, dismissively, lengthily, and unwisely. (It was at this
point that Newton sent a letter to the hunchbacked Hooke
with a mock-gracious passage about how Newton stood "on the
shoulders of giants.") Thirty years would pass—until 1704, the
year following Hooke's death—before the world would hear any
more about Newton's experiments on light.

Now, in 1686, with the first two books of the *Principia* in Halley's hands, Hooke had popped up again. For Hooke to venture yet another criticism, this time directed against Newton's

crowning work, was a sin beyond forgiving. In Newton's eyes Hooke had done nothing to contribute to a theory of gravitation. He had made a blind guess and not known how to follow it up. The challenge was not to suggest that an inverse-square law might be worth looking at, which anyone might have proposed, but to work out what the universe would look like if that law held.

Hooke had not even known how to get started, but he had airily dismissed Newton's revelations as if they were no more than the working out of a few details that Hooke had been too busy for. "Now is not this very fine?" Newton snapped. "Mathematicians that find out, settle & do all the business must content themselves with being nothing but dry calculators & drudges & another that does nothing but pretend & grasp at all things must carry away all the invention. . . ."

Hooke was a true genius, far more than Salieri to Newton's Mozart, but he did not come up to Newton's level. Hooke's misfortune was to share so many interests with a man fated to win every competition. That left both men trapped. Newton could not bear to be criticized, and Hooke could not bear to be outdone. The two men never did make peace. On the rare occasions when they found themselves thrown together, Hooke stalked out of the room. Newton was just as hostile. Even twenty years after Hooke's death, Newton could not hear his name spoken without losing his temper.

During the many years when Hooke was a dominant figure at the Royal Society, Newton made a point of staying away. When Hooke finally died, in 1703, Newton immediately accepted the post of Royal Society president. At about the same time, the Royal Society moved to new quarters. In the course of the move the only known portrait of Hooke vanished.

Chapter Forty-Nine

THE SYSTEM
OF THE WORLD

"I must now again beg you," Halley wrote Newton at the height of the Hooke affair, "not to let your resentments run so high, as to deprive us of your third book." Halley would have pleaded even more fervently if Newton had told him outright what riches he had reserved for Book III. Newton gave in to Halley's pleas. Perhaps he had meant to do so all along, although Newton seldom bothered to bark without also going on to bite.

The key to Book III was one astonishing theorem. Among the mysteries that Newton had to solve, one of the deepest was this: how could he justify the assumption that any object whatsoever, no matter how tiny or gigantic, no matter how odd its shape, no matter how complicated its makeup, could be treated mathematically as if it were a single point? Newton hadn't had a choice about simplifying things in that way, because otherwise he could not have gotten started, but it seemed an unlikely fiction.

Then, in Book III, Newton delivered an extraordinarily subtle, calculus-based proof that a complicated object could legitimately be treated as a single point. In reality the Earth was eight thousand miles in diameter and weighed thousands of billions of tons; mathematically it could be treated as a point with that same unimaginable mass. Make a calculation based on that simplifying assumption—what was the shape of the moon's orbit, say?—and the result would match snugly with reality.

Everything depended on the inverse-square law. If the universe had been governed by a different law, Newton showed, then his argument about treating objects as points would not have held, nor would the planets have fallen into stable orbits. For Newton, this was yet more evidence that God had designed the universe mathematically.

The *Principia* seemed to proclaim that message. What, after all, was the meaning of Newton's demonstration that real-life objects could be treated as idealized, abstract points? It meant that all of the mathematical arguments that Newton had made in Book I turned out to describe the actual workings of the world. Like the world's most fantastic pop-up book, the geometry text of Book I rose to life as the real-world map of Book III. Newton introduced his key findings with a trumpet flourish. "I now demonstrate the frame of the System of the World," he wrote, which was to say, "I will now lay out the structure of the universe."

And so he did. Starting with his three laws and a small number of propositions, Newton deduced all three of Kepler's laws, which dealt with the motions of the planets around the sun; he deduced Galileo's law about objects in free fall, which dealt with the motion of objects here on Earth; he explained the motion of the moon; he explained the path of comets; he explained the tides; he deduced the precise shape of the Earth.

The heart of the *Principia* was a breathtaking generalization. Galileo had made a leap from objects sliding down a ramp to objects falling through the air. Newton leaped from the Earth's pulling an apple to every pair of objects in the universe pulling one another. "There is a power of gravity," Newton wrote, "pertaining to all bodies, proportional to the several quantities of matter which they contain." *All bodies, everywhere.*

This was the theory of "universal gravitation," a single force and a single law that extended to the farthest reaches of the universe. Everything pulled on everything else, instantly and across billions of miles of empty space, the entire universe bound together in one vast, abstract web. The sun pulled the Earth, an ant tugged on the moon, stars so far away from Earth that their light takes thousands of years to reach us pull us, and we pull them. "Pick a flower on Earth," said the physicist Paul Dirac, "and you move the farthest star."

With a wave of Newton's wand, the world fell into place. The law of gravitation—one law—explained the path of a paperweight knocked off a desk, the arc of a cannonball shot across a battlefield, the orbit of a planet circling the sun or a comet on a journey that extended far, far beyond the solar system. An apple that fell a few feet to the ground, in a matter of seconds, obeyed the law of gravitation. So did a comet that traveled hundreds of millions of miles and neared the Earth only once every seventy-five years.

And Newton had done more than explain the workings of the heavens and the Earth. He had explained everything using the most familiar, literally the most down-to-earth force of all. All babies know, before they learn to talk, that a dropped rattle falls to the ground. Newton proved that if you looked at that observation with enough insight, you could deduce the workings of the cosmos.

The *Principia* made its first appearance, in a handsome,

leatherbound volume, on July 5, 1687. The scientific world searched for superlatives worthy of Newton's achievement. "Nearer the gods no mortal may approach," Halley wrote, in an adulatory poem published with the *Principia*. A century later the reverence had scarcely died down. Newton was not only the greatest of all scientists but the most fortunate, the French astronomer Lagrange declared, for there was only one universe to find, and he had found it.

Halley watched over the *Principia* all the way to the end, and past it. The Royal Society had only ventured into publishing once before. In 1685 it had published a lavish volume called *The History of Fishes* and lost money. Now the Society instructed Halley to print the *Principia* at his own expense, since he was the one who had committed it to publication in the first place. Halley agreed, though he was far from rich. The work appeared, to vast acclaim, but the Society's finances fell further into disarray. It began paying Halley his salary in unsold copies of *The History of Fishes*.

Chapter Fifty

ONLY THREE PEOPLE

From the beginning, the *Principia* had a reputation for difficulty. When Newton brushed by some students on the street one day, he heard one of them mutter, "There goes the man that writt a book that neither he nor any body else understands." It was almost true. When the *Principia* first appeared, it baffled all but the ablest scientists and mathematicians. (The first print run was tiny, between three and four hundred.) "It is doubtful," wrote the historian Charles C. Gillispie, "whether any work of comparable influence can ever have been read by so few persons."

The historian A. Rupert Hall fleshed out Gillispie's remark. Perhaps half a dozen scientists, Hall reckoned, fully grasped Newton's message on first reading it. Their astonished praise, coupled with efforts at recasting Newton's arguments, quickly drew new admirers. In time popular books would help spread Newton's message. Voltaire wrote one of the most successful, *Elements of Newton's Philosophy*, much as Bertrand Russell would later write *ABC of Relativity*. An Italian writer produced *Newtonianism for Ladies*, and an English author using the pen name Tom Telescope wrote a hit children's book.

But in physics a mystique of impenetrability only adds to a theory's allure. In 1919, when the *New York Times* ran a story on Einstein and relativity, a subheadline declared, "A Book for 12 Wise Men." A smaller headline added, "No More in All the World Could Comprehend It." A few years later a journalist asked the astronomer Arthur Eddington if it was true that only three people in the world understood general relativity. Eddington thought a moment and then replied, "I'm trying to think who the third person is."

Two features, beyond the difficulty of its mathematical arguments, made the *Principia* so hard to grasp. The first reflected Newton's hybrid status as part medieval genius, part modern scientist. Through the whole vast book Newton relies on concepts from calculus—infinitesimals, limits, straight lines homing in ever closer to curves—that he had invented two decades before. But he rarely mentions calculus explicitly or explains the strategy behind his arguments, and he makes only indirect use of calculus's labor-saving machinery.

Instead he makes modern arguments using old-fashioned tools. What looks at a glance like classical geometry turns out to be a more exotic beast, a kind of mathematical centaur. Euclid would have been baffled. "An ancient and venerable mathematical science had been pressed into service in a subject area for which it seems inappropriate," writes one modern physicist. "Newton's geometry seems to shriek and groan under the strain, but it works perfectly."

There are almost no other historical examples of so strange a performance as this use/nonuse of calculus. To get something of its flavor, we have to imagine far-fetched scenarios. Think, for instance, of a genius who grew up using Roman numerals but then invented Arabic numerals. And then imagine that he conceived an incredibly complex theory that relied heavily on the special properties of Arabic numerals—the way they make calculations easy, for instance. Finally, imagine that when he presented that theory to the world he used no Arabic numerals at all, but only Roman numerals manipulated in obscure and never-explained ways.

Decades after the *Principia*, Newton offered an explanation. In his own investigations, he said, he had used calculus. Then, out of respect for tradition and so that others could follow his reasoning, he had translated his findings into classical, geometric language. "By the help of the new Analysis [i.e., calculus] Mr. Newton found out most of the Propositions in his *Principia Philosophiae*," he wrote, referring to himself in the third person, but then he recast his mathematical arguments so that "the System of the Heavens might be founded upon good Geometry."

Newton's account made sense, and for centuries scholars took it at face value. He knew he was presenting a revolutionary theory. To declare that he had reached startling conclusions by way of a strange, new technique that he had himself invented would have been to invite trouble and doubt. One revolution at a time.

But it now turns out that Newton did *not* use calculus's shortcuts in private and then reframe them. "There is no letter," declared one of the most eminent Newtonians, I. Bernard Cohen, "no draft of a proposition, no text of any kind—not even a lone scrap of paper—that would indicate a private mode of composition other than the public one we know in the *Principia*." The reason that Newton claimed otherwise was evidently to score points against Leibniz. "He wanted," wrote Cohen, "to show that he understood and was using the calculus long before Leibniz."

This is curious, for Newton *had* understood calculus long before Leibniz, and so it would have made perfect sense for him to have drawn on its hugely powerful techniques. But he did not.

The reason, evidently, was that he was such a geometric virtuoso that he felt no impulse to deploy the powerful new arsenal that he himself had built. "As we read the *Principia*," the nineteenth-century scientist William Whewell would write, "we feel as when we are in an ancient armoury where the weapons are of gigantic size; and as we look at them, we marvel what manner of men they were who could use as weapons what we can scarcely lift as a burden."

JUST CRAZY ENOUGH

The second reason that the *Principia* was so baffling is more easily stated—the theory made no sense. This is not to deny that the theory of gravitation "works." It works astonishingly well. When NASA sent a man to the moon, every calculation along the way turned out precisely as Newton would have forecast centuries before. Nor does the model break down when applied to the farthest corners of the universe or the largest structures in nature. A theory that Newton devised by pondering the solar system and its one sun turns out to apply to galaxies made up of billions upon billions of suns, galaxies whose existence was unknown in Newton's day.

But early scientists found themselves bewildered even so. The problem was that the theory predicted, but it did not explain. Why do rocks fall? "Because of gravity," the world has said ever since Newton, but that answer only pins a name to our ignorance. Molière long ago made fun of the doctor who explained that opium makes us sleepy because it has a "dormitive potency." When Newton published the *Principia*, many scientists hailed his mathematics but denounced "gravity" as the same sort of empty explanation. They demanded to know what it meant to say that the sun pulled the planets. *How* did it pull them? What did the pulling?

Another difficulty cut deeper. Today we've grown accustomed to thinking of modern science as absurd and unfathomable, with its talk of black holes and time travel and particles that are neither here nor there. "We are all agreed that your theory is crazy," Niels Bohr, one of the twentieth century's foremost physicists, once told a colleague. "The question that divides us is whether it is crazy enough to have a chance of being correct." We think of classical science, in contrast, as a world of order and structure. But Newton's universe was as much an affront to common sense as anything that modern science has devised, and Newton's contemporaries found *his* theory crazy, too.

One of the great mysteries of modern science is where consciousness comes from. How can a three-pound hunk of gray meat improvise a poem or spin out a dream? In Newton's day, gravity was just as bewildering.* How could it be that every hunk of matter pulls every other? Newton's scheme seemed fantastically elaborate—the Alps pulled the Atlantic Ocean, which pulled back and pulled the Tower of London at the same time, which pulled Newton's pen, which pulled the Great Wall of China. How could all those pulls also stretch to the farthest corners of space, and do so instantly? How does gravity snag a comet speeding outward past the farthest planets and yank it back toward us?

---

* In time, this bewilderment died away. Darwin noted impatiently that, although his critics demanded that he explain where intelligence and awareness come from, nobody demanded a similar account of gravity. "Why is thought being a secretion of brain more wonderful than gravity a property of matter?" he asked.

Every aspect of the picture was mystifying. Gravity traveled across millions of miles of empty space? How? How could a force be transmitted with nothing to transmit it? Leibniz was only one of many eminent thinkers who hailed the brilliance of Newton's mathematics but scoffed at his physics. "He claims that a body attracts another, at whatever distance it may be," Leibniz jeered, "and that a grain of sand on earth exercises an attractive force as far as the sun, without any medium or means."

It was Newton's notion of "action at a distance" that particularly galled Leibniz and many others. Newton agreed that there was no resolving this riddle, at least for the time being, but he put it to one side. "Mysterious though it was," historian John Henry writes, by way of summarizing Newton's view, "God could make matter act at a distance—to deny this was to deny God's omnipotence."

The skeptics were not so easily satisfied. Without some mechanism that explained how physical objects pulled one another, they insisted, this new theory of universal gravitation was not a step forward but a retreat to medieval doctrines of "occult forces." Proper scientific explanations involved tangible objects physically interacting with other tangible objects, not a mysterious force that flung invisible, undetectable lassos across endless regions of space. Invoking God, said Leibniz, was not good enough. If gravity was a force that God brought about "without using any intelligible means," then that would not make sense "even if an angel, not to say God himself, should try to explain it."

Nor was the mystery merely that gravity operated across vast distances. Unlike light, say, gravity could not be blocked or affected in any way whatsoever. Hold your hand in front of your eyes and the light from a lamp on the other side of the room cannot reach you. But think of a solar eclipse. The moon passes between the Earth and the sun and blocks the sun's light, but it certainly doesn't block the gravitational force between Earth and sun—the Earth doesn't fly out of its orbit. The force seems to pass through the moon as if it weren't there.

The closer you examined Newton's theory, the more absurd it seemed. Consider, for instance, the Earth in its orbit. It travels at a fantastic speed, circling the sun at about 65,000 miles per hour. According to Newton, it is the sun's gravitational pull that keeps the Earth from flying off into space. Now imagine a giant standing atop the sun and swinging the Earth around his head at that same speed of 65,000 miles per hour. Even if the titan held the Earth with a steel cable as thick as the Earth itself, the steel would snap at once, and Earth would shoot off into the void. And yet, with no sort of cable at all, gravity holds the Earth in an unbreakable grip.

Seen that way, gravity seems incredibly powerful. But compared with nature's other forces, like electricity and magnetism, it is astonishingly feeble. If you hold a refrigerator magnet a tiny distance from the fridge, the magnet leaps through the air and sticks to the door. Which is to say, the magnetic pull of a refrigerator door outmuscles the gravitational pull of the entire Earth.

It was just as hard to understand how gravity could cut instantaneously across the cosmos. Newton maintained that it took no time whatsoever, not the briefest fraction of a second, for gravity to span even the vastest distance. If the sun suddenly exploded, one present-day physicist remarks, then according to Newton the Earth would *instantly* change in its orbit. (According to Einstein, we would be every bit as doomed, but we would have a final eight minutes of grace, blithely unaware of our fate.)

None of this made sense. Kepler and Galileo, the first great scientists of the seventeenth century, had toppled the old theories that dealt with an everyday, commonsensical world where carts grind to a halt and cannonballs fall to Earth. In its place they began to build a new, abstract work of mathematical architecture. Then Newton had come along to complete that mathematical temple.

So far, so good. Other great thinkers of the day, such men as Leibniz and Huygens, shared those mathematical ambitions. But when those peers and rivals of Newton looked closely at the *Principia*, they drew back in shock and distaste. Newton had installed at the heart of the mathematical temple not some gleaming new centerpiece, they cried, but a shrine to ancient, outmoded, occult forces.

Curiously, Newton fully shared the misgivings about gravity's workings. The idea that gravity could act across vast, empty stretches of space was, he wrote, "so great an absurdity that I believe no man who has in philosophical matters any competent faculty of thinking can ever fall into it." He returned to the point over the course of many years. "To tell us that every Species of Things is endow'd with an occult specific Quality [like gravity] by which it acts and produces manifest Effects, is to tell us nothing."

Except . . . except that the theory worked magnificently. Newton's mathematical laws gave correct answers—fantastically accurate answers—to questions that had long been out of reach, or they predicted findings that no one had ever anticipated. No one until Newton had explained the tides, or why there are two each day, or why the Earth bulges as it does, or why the moon jiggles as it orbits the Earth.

Description and prediction would have to do, then, and explanation would have to wait. How the universe managed to obey the laws he had discovered—how gravity could possibly work—Newton did not claim to know. He would not guess.

He painted himself as the voice of hardheaded reason, Leibniz as the spokesman for airy speculation. When Leibniz rebuked him for proposing so incomplete a theory, Newton maintained that restraint was only proper. He would stick to what he could know, even though Leibniz talked "as if it were a Crime to content himself with Certainties and let Uncertainties alone." Newton opted for caution. "Ye cause of gravity is what I do not pretend to know," he wrote in 1693, "& therefore would take more time to consider of it."

Twenty years later, he had made no progress. "I have not been able to discover the cause of those properties of gravity," Newton wrote in 1713, "and I frame no hypotheses." Another two centuries would pass before Albert Einstein framed a new hypothesis.

In the meantime, Newton declared his peace with his own considerable achievement. "And to us it is enough that gravity does really exist, and act according to the laws which we have explained," he wrote, in a kind of grand farewell to his theory, "and abundantly serves to account for all the motions of the celestial bodies, and of our sea."

Chapter Fifty-Two

IN SEARCH OF GOD

A different feature of Newton's theory of gravitation raised the most troubling question of all: where did God fit into Newton's universe? For seventeenth-century thinkers in general, and for Newton in particular, no question could have been more important. Today, the talk of God may seem misplaced. The *Principia* is not a sacred text but a scientific work that makes specific, quantitative predictions about the world. Those predictions are either true or false, regardless of what your religious views happen to be.* But to judge the *Principia* by the accuracy of its predictions is to see only part of it. In a similar sense, you can admire Michelangelo's *Pietà* as a gorgeous work of art even if you have no religious beliefs whatsoever. But to know what Newton thought he was doing, or Michelangelo, you need to take account of their religious motivation.

Newton had ambitions for his discoveries that stretched far beyond science. He believed that his findings were not merely

---

* The debate over whether we should look at scientists' characters and motives, or if it is only their findings that matter, continues today. "Science doesn't work because we're all nice," a NASA climatologist declared in November 2009, in the midst of a dispute over global warming. "Newton may have been an ass, but the theory of gravity still works."

technical observations but insights that could transform men's lives. The transformation he had in mind was not the usual sort. He had little interest in flying machines or labor-saving devices. Nor did he share the view, which would take hold later, that a new era of scientific investigation would put an end to superstition and set men's minds free. Newton's intent in all his work was to make men more pious and devout, more reverent in the face of God's creation. His aim was not that men rise to their feet in freedom but that they fall to their knees in awe.

So for Newton himself, the answer to the question *where does God fit in the universe?* was plain. God sat enthroned at the center of creation. Newton had always known it; he had always seen his work as a hymn to God's glory, though one written in curves and equations rather than notes on a staff. Now his dazzling success in the *Principia* provided still further evidence of the magnificence of God's design.

But the great irony of Newton's life was that many people looked at his work and drew precisely the opposite moral. Newton had not honored God, they insisted, but had made Him irrelevant. The more the universe followed laws that held everywhere and always, the less room God had to exercise his sovereignty. This critique was seldom directed at Newton personally (except by Leibniz). No one questioned the sincerity of his religious faith. Both his major works, the *Principia* and the *Opticks*, concluded with long, heartfelt outpourings of praise for the Creator. "He is eternal and infinite, omnipotent and omniscient," Newton wrote in the *Principia*. "That is, his duration reaches from eternity to eternity; his presence from infinity to infinity; he governs all things and knows all things that are or can be done."

Still, the faithful insisted, Newton had inadvertently given aid and comfort to the enemy. He had bolstered the cause of science, and science had shown itself to be an enterprise devoted to demoting God. Everyone knew that the history of religion was filled with miraculous interventions—floods, burning bushes, the sick healed, the dead returned to life. God did not merely watch his creation. On countless occasions He had stepped in and directly altered the course of events. And now, it seemed, science threatened to push God aside.

This made for a debate over gravity that in some ways anticipated the nineteenth-century battle over evolution. Such fights may seem to turn on arcane issues—planets and mathematical laws, fossils and apes—but in intellectual history, giant wars are fought on narrow battlegrounds. The real issue is always man's place in the cosmos.

Like evolution, gravity raised questions that tangled up science, politics, and theology. By hemming in God, religious thinkers railed, science promoted atheism. *Atheist*, in the seventeenth century, was an all-purpose slur that embraced a range of suspicious beliefs, much as *commie* or *pinko* would in Cold War America. But the fear it exposed was real, for to challenge religion was to call the entire social order into question. "Is nothing sacred?" was not an empty bit of rhetoric but a howl of anguish. If religion were undermined, sexual license and political anarchy were sure to follow.

Nor did science aim only at toppling age-old beliefs. Even worse, in the eyes of its detractors, the new thinkers meant to replace time-honored doctrines with their own dubious substitutes. "Scientists, like whoring Jerusalem and Babylon, have turned away from God and have put in His stead their own systems and explanations," writes one modern historian, in summarizing the antiscience case. "And these are the idols they worship: not the world created by God, but the mechanistic representations—like idols, devoid of spirit—that are the works of their own crazed imaginations."

If the universe was a machine, as science seemed to teach, then humans were just one more form of matter and there was no such thing as the soul, or choice, or responsibility. In such a world, morality would have no meaning, and, everyone would know, as one appalled writer put it, that "they may do any thing that they have a mind to."

So Newton and Leibniz squared off one last time, this time in an ideological clash over God and gravity. The battleground was the issue of God's intervention in the world. Each man accused the other of maligning God and attacking Christianity. Newton began by insisting that his theory of gravitation *did* have an explicit role for God. It was not simply that, at creation, God had set the whole solar system in motion. More than that, He had continued ever since to fine-tune His creation. The planets could not be left to run on their own, Newton's calculations showed; their ever-changing pulls on one another meant that their orbits were not quite stable. Left unattended, the solar system would fall out of kilter and, eventually, tumble into chaos.

And so the world had not been left unattended. This was, Newton maintained, still further proof of God's wisdom. If He had designed the universe to run unsupervised, He would have left room for the foolish and skeptical to argue that if God is absent *now*, perhaps He was absent *always*. God had known better.

The question of whether God had neglected his creation was so touchy that Newton's followers produced a second argument to demonstrate His ongoing presence. The miracles recorded in the Bible had taken place long ago. How to show that the age of miracles had not passed?

One way was to update the definition of *miracle*. God continued to intervene in the world, argued the theologian William Whiston, speaking on Newton's behalf, though perhaps He had changed His style. Even as familiar a feature of our lives as gravity "depends entirely on the constant and efficacious and, if you will, the supernatural and miraculous Influence of Almighty God." There was nothing inherent in the nature of rocks that caused them to fall; they fell because God *made* them fall. If you stopped to think about it, wrote Whiston, it was as miraculous for a stone to drop to the ground as it would be for it to hover in midair.

Leibniz pounced. Newton had committed heresy. Both Leibniz and Newton believed in a clockwork universe, but now Leibniz invoked the familiar image to mock his old enemy. "Sir Isaac Newton, and his followers, have also a very odd opinion concerning the work of God. According to their doctrine, God Almighty wants to wind up his watch from time to time: otherwise it would cease to move. He had not, it seems, sufficient foresight to make it a perpetual motion."

Newton fired back in fury. *He* was not the one blaspheming God. To call for a clock that ran forever on its own, as Leibniz had, was to cut God out of the picture. "If God does not concern himself in the Government of the World," declared Samuel Clarke, another of Newton's allies, ". . . it will follow that he is not an Omnipresent, All-powerful, Intelligent and Wise Being; and consequently, that he Is not at all."

Newton and Clarke were far from done. This dangerous doctrine of Leibniz's posed a threat not only to Christianity but to political stability as well. To hear Leibniz tell it, the king of the universe was a mere figurehead and not a ruler at all. Think what that meant! "If a king had a kingdom wherein all things would continually go on without his government or interposition," wrote Clarke, then he would not "deserve at all the title of king or governor." Who would need such a do-nothing king? Leibniz had allied himself with those scoundrels of whom it "may reasonably be suspected that they would like very well to set the king aside."

Leibniz did not back down. If the cosmos needed constant tinkering, as Newton would have it, then God had not fully understood a design of His own making. This was to malign God, to charge our perfect Creator with imperfection.

Deeply religious though both Newton and Leibniz were, they managed to talk past one another. The problem was that they focused on different aspects of God's greatness. Newton emphasized God's will, His ability to act however and whenever He chose. Leibniz focused on God's wisdom, His ability to see ahead of time exactly how every conceivable event would play itself out, down the furthest corridors of time.

That left both these brilliant, devout men caught in traps of their own making. Each had, in a sense, explained too much. Newton wanted above all else to portray God as a participant in the world, not a spectator. But Newton's universe seemed to run by itself, despite his protests to the contrary. That made God a kind of absentee landlord. Leibniz, on the other hand, took as *his* unbreachable principle the notion of God as all-powerful and all-knowing. The catch was that a God with those traits had no choice but to throw a switch that set in motion precisely the world we have.

The problem was that both were guilty as charged, and neither could admit it. Stuck defending indefensible positions, they fought to the death.

CONCLUSION

In the year 1600, for the crime of asserting that the Earth was one of an infinite number of planets, a man named Giordano Bruno was burned alive. Bruno, an Italian philosopher and mystic, had run afoul of the Inquisition. Charged with heresy, he was yanked from his prison cell, paraded through the streets of Rome, tied to a stake, and set afire. To ensure his silence in his last minutes, a metal spike had been driven through his tongue.

Almost exactly a century later, in 1705, the queen of England bestowed a knighthood on Isaac Newton. Among the achievements that won Newton universal admiration was this: he had convinced the world of the doctrine that had cost Giordano Bruno his life.

Sometime between those two events, at some point in the course of the 1600s, the modern world was born. Even with hindsight, pinning down the birth date is next to impossible. Still, if we who live in the new world somehow found ourselves transported to Newton's London, we would have a chance of navigating our way. In Bruno's Rome we would founder and drown. And since those earliest days, the pace of change has only accelerated. The world has raced ahead, permanently in fast-forward, with science and technology taking an ever more conspicuous spot in the foreground.

In the decades following his death, Newton's reputation continued to soar. Though gravity remained as mysterious as ever, new generations of scientists built on Newton's theories to produce an ever more detailed, ever more accurate picture of the universe. Each step forward provided still more proof that Newton had read God's mind.

Perhaps the most dramatic confirmation came in 1846, when a French mathematician named Urbain Le Verrier looked hard at Newton's laws, sat down to calculate, and discovered a new planet. This was Neptune, discovered by deduction. Le Verrier and other astronomers of the day knew that the orbit of the planet Uranus was not exactly what theory predicted. The reason, they proposed, was that some unseen planet was tugging it off course. Using Newton's laws, Le Verrier managed to calculate the vital statistics—the mass, position, and path—of this supposed planet. He sent his results to the German astronomer Joseph Galle. Le Verrier's letter reached Galle on September 23, 1846. On the same evening, Galle directed his telescope to the spot in the sky that Le Verrier had identified. There, just barely visible, he found Neptune.

Long before Le Verrier, the successes racked up by Newton's followers had inspired the hope of similar breakthroughs in *every* field. Just as Newton had discovered the laws of inanimate nature, so would some new thinker find the laws of human nature. A handful of rules would explain all the apparent happenstance of history, psychology, and politics. Better still, once its laws came to be understood, society could be reshaped in a rational way.

America's founding fathers argued explicitly that the success of the scientific approach foretold their own success. Free minds would make the world anew. Rather than defer to tradition and authority, the new thinkers would start from first principles and build on that sturdy foundation. Kings and other accidental tyrants would be overthrown, sensible and self-regulating institutions set in their place. In the portrait of himself that he liked best, Benjamin Franklin sat deep in thought in front of a bust of Newton, who watched his protégé approvingly. Thomas Jefferson installed a portrait of Newton in a place of honor at Monticello.

As they spelled out the design of America's political institutions, the founders clung to the model of a smooth-running, self-regulating universe. In the eyes of the men who made America, the checks and balances that ensured political stability were directly analogous to the natural pushes and pulls that kept the solar system in balance. "The Constitution of the United States had been made under the dominion of the Newtonian theory," Woodrow Wilson would later write. If you read the *Federalist* papers, Wilson continued, the evidence jumped out "on every page." The Constitution was akin to a scientific theory, and the amendments played the role of experiments that helped define and test that theory.

Newton's posthumous influence was overwhelming, but in one respect his triumph proved *too* complete. Newton would have wept with rage to know that his scientific descendants spent their lifetimes proving conclusively that the clockwork universe ran even more smoothly than he had ever believed. It ran so marvelously well, in fact, that a new consensus quickly arose—just as Newton's enemies had claimed, Newton *had* built a universe that had no place within it for God.

The crowning glory of eighteenth-century astronomy was the proof, by the French mathematician Pierre Simon Laplace, that although the planets did wobble a bit as they circled the sun, those wobbles stayed within a narrow, predictable range. Since the wobbles did not grow larger and larger as time passed, as Newton had believed, they did not require that God step in to smooth things out. Laplace presented his masterpiece, a tome called *Celestial Mechanics*, to Napoleon.

How was it, Napoleon asked, that in all those hundreds of pages, Laplace had made not a single mention of God?

"I had no need of that hypothesis," Laplace told the emperor.

Newton outlived his longtime enemy Leibniz. "Mr. Leibniz is dead, and the dispute is finished," a colleague wrote Newton in 1716. It was not finished; even without an enemy, Newton fought on for another six years. For a long while, posterity would treat Leibniz with scarcely more regard. Newton's achievements were celebrated by the likes of Alexander Pope and William Wordsworth, who composed worshipful verses in his honor. Leibniz had the misfortune to stir the wrath of Voltaire, the greatest wit of his age, who caricatured him in a book still read today.

At least in scientific circles, though, Leibniz's reputation has grown through the centuries. In every history of logic or computers, especially, his ahead-of-their-age insights now meet with stunned admiration. Even in physics, where his ideas have long since been abandoned, his ambitious dreams still thrive. Today's physicists toss around such phrases as a "theory of everything." Leibniz would have felt right at home.

Near the end, Leibniz had received a letter from Caroline, Princess of Wales, his onetime pupil. She sent word that the king might possibly, at last, bring him to England. "Nothing could give me a greater desire to go there than the kindnesses of Your Royal Highness," Leibniz wrote back, "but as I do not hope to go soon, I do not know if I can hope to go later; for there is not a lot of later to hope for in me."

Leibniz died in Germany, neglected, nearly alone, and beset by a host of painful ailments. He was buried in an unmarked grave (a marker was eventually added). "You would have thought it was a felon they were burying," wrote one of the few funeral guests, "instead of a man who had been an ornament to his country."

Newton's body lies in Westminster Abbey beneath a marble statue. Perhaps it is fitting that Newton was treated as nearly godlike and Leibniz as merely mortal. "The more I got to know Leibniz," one recent biographer wrote, "the more he seemed to me all-too-human, and I quarreled with him." No one ever directed the same complaint against Newton. Leibniz was too human, and Newton seemed scarcely human at all.

In the 1980s, as we have seen, the astrophysicist Subrahmanyan Chandrasekhar, a scientist of towering reputation, went through the *Principia* line by line in an attempt to probe the mind of his predecessor. "During the past year," Chandrasekhar told me in a 1987 interview, "I've taken proposition after proposition, written out my own proof, and then compared it with Newton's. In every case, his proofs are incredibly concise; there is not a superfluous word. The style is imperial, just written down as if the insights came from Olympus."

"If you take great scientists," Chandrasekhar went on, "even though they made discoveries that one could not have made oneself, one can *imagine* making them—people say, 'I could have done that, but I was just stupid.' Normal scientists can think of greater men, and it is not difficult to imagine doing what they did. But I don't think it's possible for any scientist to imagine what it would have been like to be Newton."

Temperamentally, the gulf was nearly as big as it was intellectually. The usual consolations of life, friendship and sex included, appealed to Newton hardly at all. Art, literature, and music had scarcely more allure. He dismissed the classical sculptures in the Earl of Pembroke's renowned collection as "stone dolls." He waved poetry aside as "a kind of ingenious nonsense." He rejected opera after a single encounter. "The first Act I heard with pleasure, the 2d stretch'd my patience, at the 3d I ran away."

"If we evolved a race of Isaac Newtons, that would not be progress," Aldous Huxley once remarked, with a mix of wonder and horror. "For the price Newton had to pay for being a supreme intellect was that he was incapable of friendship, love, fatherhood, and many other desirable things. As a man he was a failure; as a monster he was superb."

Huxley's notion of a Faustian trade-off smacks a bit of sour grapes, as if to reassure the rest of us that this genius business is not all we imagine it to be. But Huxley was right to emphasize the gulf between Newton and everyone else. Newton's best biographer, Richard Westfall, told me many years ago that, in the course of examining Newton's life, he had lived with the man for twenty years. Westfall's magnum opus, *Never at Rest*, is a model of insight and empathy, but Westfall lamented that he never felt he knew Newton. On the contrary, Newton came to seem ever more mysterious, not only in intellect but in motives and hopes, fears and ambitions. "The more I learned," Westfall recalled, "the more I realized how far he was from me, in every regard." Newton was, Westfall declared, "wholly other."

Newton's contemporaries sensed the same gap. When the *Principia* was new, the Marquis de L'Hôpital, a skilled mathematician, read it with incredulity. L'Hôpital had been pondering a technical question about how streamlined objects move through fluids, and an English mathematician showed him that Newton had worked out a solution in the *Principia*. "He cried out with admiration Good god what a fund of knowledge there is in that book? he then asked the Doctor every particular about Sir Isaac even to the colour of his hair said does he eat & drink & sleep. Is he like other men?"

In all the important ways, he was not like other men. Perhaps we would do better to acknowledge the gulf than to try to bridge it. At Cambridge, Newton could occasionally be seen standing in the courtyard, staring at the ground, drawing diagrams in the gravel with a stick. Eventually he would retreat indoors. His fellow professors did not know what the lines represented, but they stepped carefully around them, in order to avoid hindering the work of the lonely genius struggling to decipher God's codebook.

ACKNOWLEDGMENTS

My first career ambition, years ago, was to play professional basketball. This plan did not long endure. It was succeeded by a far longer-lived but perhaps equally foolish notion, to spend a lifetime studying theoretical mathematics. After several years wandering dazed through infinite dimensional spaces, I left the hunt to those better suited to it. But I owe thanks to a host of mentors, Fred Solomon and Gene Dolnick notable among them, who first opened my eyes to mathematical beauty.

In researching this book I pestered many long-suffering physicists, historians, and philosophers with queries about everything from spiral galaxies to Leibniz's thoughts on unicorns. I owe special gratitude to Rebecca Grossman, Mike Briley, Cole Miller, and, especially, Larry Carlin, who carried out, solely for my benefit, the best of all possible philosophy tutorials. Steven Shapin, an eminent historian of science, generously shared his deep insights into science and the 1600s. Owen Gingerich and Simon Schaffer sorted out historical mysteries that had stumped me. All my guides have been disabused of the belief that there's no such thing as a foolish question.

Michele Missner once again tracked down countless articles, the more obscure the better. Katerina Barry, an unflappable researcher as well as an artist and web designer, gathered images from libraries and museums across Europe and America. Rob Crawford resolved crises large and small with skill and grace. Hugh Van Dusen, my friend and my editor, demonstrated once again that he is the ideal ally.

My sons Sam and Ben, both writers, read every draft and weighed in on every editorial decision. No one could have better colleagues.

Lynn deserves more thanks than I know how to put in words.

NOTES

Sources for quotations and for assertions that might prove elusive can be found below. To keep these notes in bounds, I have not documented facts that can be readily checked in standard sources. Publication information is provided in the notes only for those sources not listed in the bibliography.

EPIGRAPH

The universe is but a watch: Bernard de Fontenelle, *Conversations on the Plurality of Worlds* (London, 1803), p. 10.

PREFACE

xv *The murder rate*: Manuel Eisner, "Modernization, Self-Control, and Lethal Violence. The Long-Term Dynamics of European Homicide Rates in Theoretical Perspective," *British Journal of Criminology* 41, no. 4 (2001).

xv *"a sooty Crust or Furr"*: Barbara Freese, *Coal: A Human History* (New York: Penguin, 2004), p. 35, quoting John Evelyn.

xv *"a stinking, muddy, filth-bespattered"*: J. H. Plumb, *The First Four Georges* (London: Fontana, 1981), p. 17.

xvi *The same barges*: Emily Cockayne, *Hubbub: Filth, Noise, and Stench in England*, p. 93.

xvi *When Shakespeare and his fellow*: Gregory Clark, *A Farewell to Alms: A Brief Economic History of the World* (Princeton, NJ: Princeton University Press, 2007), p. 107.

xvi *the palace at Versailles*: Katherine Ashenburg, *The Dirt on Clean*, p. 116.

xvifn *The historian Jules Michelet*: Ashenburg, *The Dirt on Clean*, p. 12. Ashenburg notes that Michelet exaggerated. She puts the correct figure at four centuries.

xvii *"Men expected the sun"*: Alfred North Whitehead, *Science and the Modern World*, p. 5.

CHAPTER 1. LONDON, 1660

4 *skeletally thin Robert Boyle*: Steven Shapin, *A Social History of Truth*. Shapin devotes a fascinating chapter to the riddle of "Who Was Robert Boyle?"

4 *Boyle maintained* three: Lisa Jardine, *On a Grander Scale*, p. 194.

4 *"low of stature"*: Leo Hollis, *London Rising*, p. 48.

4 *a "miracle of youth"*: Jardine, *On a Grander Scale*, p. 236, quoting John Evelyn.

5 *"the most fearful"*: John Maynard Keynes, "Newton, the Man," p. 278, quoting the Cambridge mathematician William Whiston.

CHAPTER 2. SATAN'S CLAWS

7 *"Any cold might be"*: Peter Earle, *The Making of the English Middle Class: Business, Society and Family Life in London 1660–1730* (Berkeley: University of California Press, 1989), p. 302.

7 *life expectancy was only*: Keith Thomas, *Religion and the Decline of Magic*, p. 5.

7 *London was so disease-ridden*: A. Lloyd Moote and Dorothy Moote, *The Great Plague*, p. 26.

7 *"puppy boiled up"*: Anna Beer, *Milton*, p. 386.

8 *"I have had the misfortune"*: Earle, *The Making of the English Middle Class*, p. 302.

8 *When Charles II suffered*: T. B. Macaulay, *History of England*, ch. 4, "James the Second," available at http://www.strecorsoc.org/macaulay/m04a.html. I drew details from Macaulay's *History*; Antonia Fraser's *Royal Charles*, p. 446; and an account by the king's chief physician, Sir Charles Scarburgh, at http://tinyurl.com/y3wgtom.

8 *"For what is the cause"*: Adam Nicolson, *God's Secretaries* (New York: Harper, 2005), p. 25.

10 *"People lived in continual terror"*: Morris Kline, *Mathematics in Western Culture*, p. 235.

11 *"Those are my best days"*: Eugen Weber, *Apocalypses*, p. 100.

12 *"Threatening my father and mother"*: Richard Westfall includes the entire list in his "Writing and the State of Newton's Conscience."

12 *writer and theologian Isaac Watts*: Roy Porter, *The Creation of the Modern World*, p. 157.

CHAPTER 3. THE END OF THE WORLD

13 *"The trumpet would sound"*: Perry Miller, "The End of the World," p. 171.

14 *"Books on the Second Coming"*: Frank Manuel, *A Portrait of Isaac Newton*, p. 129.

14fn *Christopher Wren's father*: Adrian Tinniswood, *His Invention So Fertile: A Life of Christopher Wren*, p. 17.

14 *"great apostasy"*: Richard Westfall, *Never at Rest*, p. 321.

15 *"What shall be the sign"*: Matthew 24:3, King James Bible.

15 *"sexual musical chairs"*: Lawrence Stone, *The Family, Sex, and Marriage*, p. 328.

16 *"So horrible was it"*: *David Levy's Guide to Observing and Discovering Comets* (New York: Cambridge University Press, 2003), p. 9, quoting Ambroise Pare.

16 *"The thick smoke"*: Tinniswood, *His Invention So Fertile*, p. 10, quoting Andreas Celichius.

17 *"a Coffin," floating*: Ibid., p. 11.

17 *"this comet portends pestiferous"*: Moote and Moote, *The Great Plague*, p. 20.

17 *clouds of flies*: J. Fitzgerald Molloy, *Royalty Restored* (London: Downey, 1897), p. 167.

17 *"A deformed monster"*: Neil Hanson, *The Great Fire of London*, p. 28.

18 *Robert Boyle, renowned today*: Westfall, *Science and Religion in Seventeenth-Century England*, p. 124. The historian Frank Manuel discusses Boyle's belief in the imminence of the apocalypse in *Portrait of Isaac Newton*, p. 129.

18 *"The fourth beast [in the book of Revelation]"*: Isaac Newton, *Observations upon the Prophecies of Daniel, and the Apocalypse of St. John*, part 1, ch. 4, "Of the vision of the four Beasts." This posthumous work by Newton can be found, along with seemingly everything else Newton-related, at the indispensable Newton Project website, http://www.newtonproject.sussex.ac.uk/prism.php?id=1. This essay is at http://www.newtonproject.sussex.ac.uk/view/texts/normalized/THEM00198.

CHAPTER 4. "WHEN SPOTTED DEATH RAN ARM'D THROUGH EVERY STREET"

20 *"like cheese between layers"*: Norman Cantor, *In the Wake of the Plague*, p. 8.

20 *"Oh happy posterity"*: Barbara Tuchman, *A Distant Mirror* (New York: Ballantine, 1978), p. 99.

21 *"Great fears of the sicknesse"*: Samuel Pepys's diary entry for April 30, 1665, available at www.pepysdiary.com.

22 *"A nimble executioner"*: Margaret Healy, "Defoe's *Journal* and the English Plague Writing Tradition," quoting the seventeenth-century pamphleteer Thomas Dekker.

22 *"the surest Signes"*: This quote and the description of plague symptoms in the next several sentences come from Richelle Munkhoff, "Searchers of

the Dead: Authority, Marginality, and the Interpretation of Plague in England, 1574–1665," *Gender and History* 11, no. 1 (April 1999).

23 *despised old women called "searchers"*: Ibid.

24 *"Death was the sure midwife"*: Nathaniel Hodge, *Loimolgia, or An Historical Account of the Plague in London in 1665.* See http://rbsche.people.wm.edu/ H111_doc_loimolgia.html.

24 *"Poor Will that used to sell"*: Pepys's diary, August 8, 1665.

CHAPTER 5. MELANCHOLY STREETS

25 *"Multitude of Rogues"*: Roger Lund, "Infectious Wit: Metaphor, Atheism, and the Plague in Eighteenth-Century London," *Literature and Medicine* 22, no. 1 (Spring 2003), p. 51.

25 *kill "all their dogs"*: Moote and Moote, *The Great Plague*, p. 177.

26 *"when we have purged"*: Tinniswood, *His Invention So Fertile*, p. 115, quoting Henry Oldenburg, secretary of the Royal Society.

26 *"Little noise heard day or night"*: Letter written September 4, 1664, by Pepys to Lady Carteret, in *Correspondence of Samuel Pepys*, vol. 5, p. 286. See http://tinyurl.com/y2aqoze.

27 *"A just God now visits"*: John Kelly, *The Great Mortality*, p. xv.

27 *"But Lord, how empty"*: Pepys's diary, October 16, 1665.

28 *Builders would one day*: Raymond Williamson, "The Plague in Cambridge," *Medical History* 1, no. 1 (January 1957), p. 51.

CHAPTER 6. FIRE

29 *Iron bars in prison cells*: Hanson, *The Great Fire of London*, p. 165, quoting John Evelyn.

30 *what Robert Boyle called*: Moote and Moote, *The Great Plague*, p. 69.

30 *"Pish!" he said*: Christopher Hibbert, *London* (London: Penguin, 1977), p. 67, and Hanson, *The Great Fire of London*, p. 49.

31 *Even on the opposite sides*: G. M. Trevelyan, *English Social History* (New York: Penguin, 1967), p. 305.

32 *Slung over his shoulder*: Antonia Fraser, *Royal Charles*, p. 245.

32 *"A horrid noise the flames made"*: Pepys's diary, September 2, 1666.

32 *Stones from church walls exploded*: Hollis, *London Rising*, p. 121.

32 *"God grant mine eyes"*: John Evelyn, *The Diary of John Evelyn*, vol. 2, p. 12. This is from Evelyn's diary entry for September 3, 1666, available at http./ www.pepysdiary.com/indepth/archive/2009/09/02/evelyns–fire.php.

33 *People wandered in search*: Hollis, *London Rising*, p. 122.

33 *"The ground was so hot"*: Hanson, *The Great Fire of London*, p. 163.

33 *"Now nettles are growing"*: Ibid., p. xv, quoting from a pamphlet by Thomas Vincent, *God's Terrible Voice in the City*.

CHAPTER 7. GOD AT HIS DRAWING TABLE

35 *God had fashioned the best*: Philosophers still debate precisely how Leibniz reconciled his belief that God had created the best possible world with his (apparent) belief in a day of judgment. One notion is that divine punishment was a feature of even the best possible world, because harmony required both that virtue be rewarded and sin punished.

36 *Newton and many of his peers*: J. E. McGuire and P. M. Rattansi, "Newton and the 'Pipes of Pan,'" p. 135. See also Piyo Rattansi, "Newton and the Wisdom of the Ancients," in John Fauvel et al., eds., *Let Newton Be!*, p. 187; Force and Popkin, *Newton and Religion*, p. xvi; Steven Shapin, *The Scientific Revolution*, p. 74.

37 *By far the most important*: The only challenges to the mainstream view came from the much-feared, much-reviled Thomas Hobbes and Baruch Spinoza.

37 *"All disorder," wrote Alexander Pope*: Pope, "An Essay on Man."

37fn *"this continued sterility"*: Jane Dunn, *Elizabeth and Mary* (New York: Vintage, 2005), p. 17.

38 *"too paganish a word"*: Thomas, *Religion and the Decline of Magic*, p. 79.

38 *The very plants in the garden*: Lorraine Daston and Katharine Park, *Wonders and the Order of Nature, 1150–1750*, p. 296, quoting Walter Charleton, *The Darkness of Atheism Dispelled by the Light of Nature*.

38 *"People rarely thought of themselves"*: Jacques Barzun, *From Dawn to Decadence*, p. 24.

38 *Atheism was literally unthinkable*: People called their enemies "atheists," but the charge had to do with behaving badly—acting in ways that offended God—rather than with denying God's existence. *Atheist* was a catch-all slur directed at the immoral and self-indulgent.

38 *Even Blaise Pascal*: Arthur Lovejoy, *The Great Chain of Being*, p.153.

40 *Plato proposed that a free man*: Morris Kline, *Mathematics: The Loss of Certainty*, p. 22. See Plato's *Laws*, book 11. "He who in any way shares in the illiberality of retail trades may be indicted for dishonouring his race by any one who likes . . . and if he appear to throw dirt upon his father's house by an unworthy occupation, let him be imprisoned for a year and abstain from that sort of thing; and if he repeat the offence, for two years; and every time that he is convicted let the length of his imprisonment be doubled." See http://classics.mit.edu/Plato/laws.11.xi.html.

41 *"It is ye perfection"*: Richard Westfall, *Never at Rest*, p. 327.

CHAPTER 8. THE IDEA THAT UNLOCKED THE WORLD

42 *"decipher the page and chapter"*: John Carey, *John Donne*, p. 128.

42 *the mysteries of multiplication*: Pepys's diary, July 4, 1662.

43 *a mathematics of change*: Ernst Cassirer, "Newton and Leibniz," p.381. See also Karen Armstrong, *A History of God*, p. 35.

44 *"the most truly revolutionary"*: I. Bernard Cohen, *Revolution in Science*, p. 90.

44 *a widow, not yet thirty*: Hannah Newton's birth date is unknown. The Newton biographer Frank Manuel suggests that she was probably around thirty when she married for the second time, three years after Isaac's birth. See Manuel, *A Portrait of Isaac Newton*, p. 24.

45 *"When one . . . compares"*: Matthew Stewart, *The Courtier and the Heretic*, p. 12.

45 *Frederick the Great declared*: Daniel Boorstin, *The Discoverers*, p. 414.

46 *"I invariably took"*: Stewart, *The Courtier and the Heretic*, p. 43.

46 *His favorite wedding gift*: Bertrand Russell, *A History of Western Philosophy* (New York: Simon & Schuster, 1945), p. 582.

46 *slept in his clothes*: Gale Christianson, *Isaac Newton*, p. 65.

46 *seventeen portraits*: Peter Ackroyd, *Newton*, p. 98.

46 *so much time working with mercury*: Milo Keynes, "The Personality of Isaac Newton," p. 27.

46 *"It's so rare," the Duchess*: Stewart, *The Courtier and the Heretic*, p. 12.

47 *"a Machine for walking on water"*: The drawing comes from a 1637 text by Daniel Schwenter, a German mathematician and inventor, titled *Deliciae physico-mathematicae*. Leibniz witnessed (and was much impressed by) a similar demonstration several decades later.

47 *"To remain fixed in one place"*: Ibid., p. 53.

47 *walk on water*: Philip Wiener, "Leibniz's Project," p. 234.

48 *"his cat grew very fat"*: Westfall, *Never at Rest*, p. 103.

48 *"His peculiar gift"*: John Maynard Keynes, "Newton, the Man," p. 278.

49 *"I took a bodkin"*: Westfall, *Never at Rest*, p. 94.

CHAPTER 9. EUCLID AND UNICORNS

50 *"weapon salve"*: Liza Picard, *Restoration London*, p. 78.

50 *"a living chameleon"*: Charles Richard Weld, *History of the Royal Society* (London: John W. Parker, 1848), v. 1, p. 114.

51 *The spider, unfazed*: Ibid., p. 113.

51 *Newton's paper followed*: Robert Crease, *The Prism and the Pendulum*, p. 72.

51 *Visitors ogled such marvels*: Christopher Hibbert, *London*, p. 100.

51 *the best cure for cataracts*: Stone, *The Family, Sex, and Marriage*, p. 65.

52 *a "flying chariot"*: Marjorie Nicolson and Nora Mohler, "Swift's 'Flying Island' in the *Voyage to Laputa*," p. 422.

52 *Since reliable men vouched*: John Henry, "Occult Qualities and the Experimental Philosophy," p. 359. The highly regarded member of the Royal Society was Joseph Glanvill.

52 *the seas contained mermaids*: John Locke, *An Essay Concerning Human Understanding*, book 3, ch. 6, "Of the Names of Substances" (London: Thomas Tegg, 1841), p. 315.

52 *an ancient National Enquirer*: Daston and Park, *Wonders and the Order of Nature*, p. 231.

53 *The "Tyburn tree"*: The gallows stood near what is now Speakers' Corner in Hyde Park.

53 *"All the Way, from Newgate to Tyburn"*: Simon Devereaux, "Recasting the Theater of Execution," *Past & Present* 202, no. 1 (February 2009).

53 *a hand's "death sweat"*: Hanson, *The Great Fire of London*, p. 216, and Thomas, *Religion and the Decline of Magic*, p. 204.

54 *the corpse "identified"*: Daston and Park, *Wonders and the Order of Nature*, p. 241.

54 *painstakingly dissected one witch's*: Thomas, *Religion and the Decline of Magic*, p. 644.

55 *the "philosopher's stone"*: Christianson, *Isaac Newton*, p. 55.

55 *some half million words*: Rattansi, "Newton and the Wisdom of the Ancients," p. 193.

55 *Leibniz's only fear*: Stewart, *The Courtier and the Heretic*, p. 48.

55 *"Whatever his aim"*: Christianson, *Isaac Newton*, p. 55.

55 *He never spoke of*: Westfall, *Never at Rest*, p. 298.

55 *"the Green Lion"*: William Newman, Indiana University historian of science, speaking on PBS in a *Nova* program, *Newton's Dark Secrets*, broadcast November 15, 2005.

56 *"Just as the world was created"*: Jan Golinski, "The Secret Life of an Alchemist," in *Let Newton Be!*, p. 160.

56 *Keynes purchased a trove*: For an excellent, detailed history of Newton's papers, see http://www.newtonproject.sussex.ac.uk/prism.php?id=23.

56 *"the last of the Babylonians"*: John Maynard Keynes, "Newton, the Man," p. 277.

CHAPTER 10. THE BOYS' CLUB

58 *New arrivals found places*: Tinniswood, *His Invention So Fertile*, p. 79.

59 *"the expansive forces"*: Marjorie Nicolson and Nora Mohler, "The Scientific

Background of Swift's *Voyage to Laputa*," in Nicolson, *Science and Imagination*, p. 328.

59 "*We put in a snake*": Lisa Jardine, *Ingenious Pursuits*, p. 114.

60 "*A man thrusting in his arm*": Lisa Jardine, *The Curious Life of Robert Hooke*, p. 105. Jardine writes that the unnamed showman with his arm in the pump was "almost certainly Hooke."

60 *one Arthur Coga*: Weld, *History of the Royal Society*, vol. 1, p. 220.

60 *a perfect subject*: I owe this insight to Steven Shapin, "The House of Experiment in Seventeenth-Century England," p. 376.

61 "*to be the Author of new things*": Boorstin, *The Discoverers*, p. 409, quoting Thomas Sprat, *History of the Royal Society*, (London: 1734), p. 322.

62 "*old wood to burn*": Cohen, *Revolution in Science*, p. 87.

62 "*not to discover the new*": Boorstin, *The Discoverers*, p. 409.

62 *In the fourteenth century Oxford*: John Barrow, *Pi in the Sky* (New York: Oxford University Press, 1992), p. 205.

63 "*the hallmark of the narrow-minded*": Daston and Park, *Wonders and the Order of Nature*, p. 61.

63 "*For God is certainly called*": Ibid., p. 39.

63 The "*lust to find out*": William Eamon, *Science and the Secrets of Nature*, p. 60.

63 "*what the Lord keeps secret*": Ecclesiastes 3:22–23, quoted in Eamon, *Science and the Secrets of Nature*, p. 60.

64 "*If the wisest men*": Westfall, *Science and Religion in Seventeenth-Century England*, p. 22.

64 *How could anyone draw*: Shapin, *The Scientific Revolution*, p. 82.

64 "*absorbing, classifying, and preserving*": Allan Chapman, *England's Leonardo: Robert Hooke and the Seventeenth-Century Scientific Revolution*, p. 40.

65fn *Bacon's zeal for experimentation*: John Aubrey, *Brief Lives* (Woodbridge, Suffolk: Boydell, 1982), entry for "Francis Bacon."

65 *Nature must be "put to the torture"*: Ibid., p. 40.

65 *dizzy and temporarily deaf*: Jardine, *Ingenious Pursuits*, p. 56.

CHAPTER 11. TO THE BARRICADES!

66 "*I swear to you by God's*": David Berlinski, *Infinite Ascent*, p. 66.

67 "*like torches, that in the lighting*": Eamon, *Science and the Secrets of Nature*, p. 330.

68 "*I've known since yesterday*": Simon Singh, *Big Bang* (New York: Harper, 2004), p. 302. Richard Feynman tells the story in its classic, romantic form in his *Feynman Lectures on Physics* (Reading, MA: Addison-Wesley, 1963), pp. 3–7, almost as soon as he begins.

68   *For decades Hooke argued*: Eamon, *Science and the Secrets of Nature*, p. 347.

68   *"Nothing considerable in that kind"*: Ibid., p. 347.

68   *"Do not throw your pearls"*: Paolo Rossi, *The Birth of Modern Science*, p. 18.

69fn *The historian Paolo Rossi*: Rossi, *The Birth of Modern Science*, p. 15.

70   *"to improve the knowledge"*: Eamon, *Science and the Secrets of Nature*, p. 348.

70   *"not by a glorious pomp"*: Ibid., p. 25, quoting Sprat, *History of the Royal Society*, pp. 62–63.

70   *"a close, naked, natural way"*: Sprat, *History of the Royal Society*, p. 113.

71   *"All that I mean"*: Carey, *John Donne*, p. 58.

CHAPTER 12. DOGS AND RASCALS

72   *"If you would like"*: Rossi, *The Birth of Modern Science*, p. 24.

73   *"glacial remoteness"*: The modern physicist is Subrahmanyan Chandrasekhar; the remark comes from a talk he gave in April 1975 at the University of Chicago, titled "Shakespeare, Newton, and Beethoven, or Patterns of Creativity," available at http://www.sawf.org/newedit/edit02192001/musicarts.asp.

73   *Samuel Johnson's remark*: James Boswell, *Life of Johnson* (London: Henry Frowde, 1904), vol. 2, p. 566.

73fn *The esteemed eighteenth-century*: Laplace's despairing admirer was Nathaniel Bowditch, quoted in Dirk Struik, *A Concise History of Mathematics*, p. 135.

73   *"baited by little Smatterers"*: Westfall, *Never at Rest*, p. 459.

74   *"the first time that a major"*: Merton, *On the Shoulders of Giants*, p. 11, quoting I. Bernard Cohen, *Franklin and Newton*.

75   *Hooke denounced his enemies*: Steven Shapin, "Rough Trade," *London Review of Books*, March 6, 2003, reviewing *The Man Who Knew Too Much: The Strange and Inventive Life of Robert Hooke*, by Stephen Inwood.

75   *Newton's aim was evidently*: Manuel, *A Portrait of Isaac Newton*, p. 145, and Mordechai Feingold, *The Newtonian Moment*, pp. 23–24.

CHAPTER 13. A DOSE OF POISON

76   *dissections had been performed*: Terence Hawkes, *London Review of Books*, December 11, 1997, reviewing *Issues of Death: Mortality and Identity in English Renaissance Tragedy*, by Michael Neill.

76   *"the culture's preference"*: Ibid.

77   *the Quaker James Nayler*: Beer, *Milton*, p. 301.

77   *a section titled "Excursions"*: Picard, *Restoration London*. Pepys certainly thought of executions in this casual way. On October 13, 1660, he found himself with some unexpected free time. "I went out to Charing Cross,"

Pepys wrote in his diary, "to see Major-general Harrison hanged, drawn, and quartered; which was done there, he looking as cheerful as any man could do in that condition. He was presently cut down, and his head and heart shown to the people, at which there was great shouts of joy." Within a sentence or two, Pepys went on to report that he'd eaten oysters for dinner.

77 *"A man sentenced to this terrible"*: Picard, *Restoration London*, p. 211.

77 *To preserve severed heads*: Beer, *Milton*, p. 302.

78 *traitors' heads impaled on spikes*: The account of London Bridge (and the reference to Thomas More) comes from Picard, *Restoration London*, p. 23. See also Aubrey, *Brief Lives*, "Sir Thomas More," and Paul Hentzner, *Travels in England During the Reign of Queen Elizabeth*, available at http://ebooks. adelaide.edu.au/h/hentzner/paul/travels/chapter1.html.

79 *"Whatever others think"*: Pepys's diary entry for February 17, 1663.

79 *Newton veered toward vegetarianism*: Steven Shapin, "Vegetable Love," *New Yorker*, January 22, 2007, reviewing *The Bloodless Revolution: A Cultural History of Vegetarianism from 1600 to Modern Times*, by Tristram Stuart.

80 *"The result was a melody"*: Thomas Hankins and Robert Silverman, *Instruments and the Imagination* (Princeton, NJ: Princeton University Press, 1999), pp. 73, 247.

80 *"that traditional nursery rhymes portray"*: Keith Thomas, *Man and the Natural World*, p. 147.

80 *With a dog tied*: Tinniswood, *His Invention So Fertile*, p. 1.

80 *Boyle subjected his pet setter*: Ibid., p. 34.

80fn *The word* disease: Moote and Moote, *The Great Plague*, p. 141.

81 *Boyle wrote a paper*: Robert Boyle, "Trial proposed to be made for the Improvement of the Experiment of Transfusing Blood out of one Live Animal into Another," *Philosophical Transactions*, February 11, 1666, available at http://rstl.royalsocietypublishing.org/content/1/1-22/385.

81 *"a foreign Ambassador"*: The ambassador intended to test the effects of a substance called Crocus metallorum, sometimes used as a medicine to induce vomiting.

81 *The servant spoiled*: Tinniswood, *His Invention So Fertile*, p. 37.

82 *"The first died upon the place"*: Pepys's diary, November 14 and 16, 1666.

CHAPTER 14. OF MITES AND MEN

83 *he set up a borrowed telescope*: Claire Tomalin, *Samuel Pepys*, p. 248.

83 *he raced out to buy a microscope*: Pepys's diary, August 13, 1664.

83 *he struggled through Boyle's*: Pepys's diary, June 4, 1667.

83 *"a most excellent book"*: Pepys's diary, June 10, 1667.

83fn *Like James Thurber*: Thurber described his attempts to master the microscope in *My Life and Hard Times*.

84 *his "jesters"*: Michael Hunter, *Science and Society in Restoration England*, p. 131. See also Pepys's diary, February 1, 1664.

84 *"Ingenious men and have found out"*: Hunter, *Science and Society in Restoration England*, pp. 91–92.

84 *"I shall not dare to think"*: Hunter, *Science and Society in Restoration England*, pp. 91–92.

84 *"Should those Heroes go on"*: Manuel, *A Portrait of Isaac Newton*, p. 130, quoting Joseph Glanvill. Glanvill's remark is from his *Vanity of Dogmatizing*, written in 1661.

86 *Gimcrack studied the moon*: Claude Lloyd, "Shadwell and the Virtuosi." The Shadwell quotes come from Lloyd's essay.

86 *Hooke went to see the play*: Shapin, "Rough Trade."

86 *Samuel Butler lampooned*: In his poem *Hudibras*, part 2, canto 3.

87 *Swift visited the Royal Society*: Nicolson and Mohler, "The Scientific Background of Swift's *Voyage to Laputa*," p. 320.

87 *"softening Marble for Pillows"*: Jonathan Swift, *Gulliver's Travels*, part 3, ch. 5.

87 *"one Man shall do the Work"*: Ibid., part 3, ch. 4.

88 *"a Shoulder of Mutton"*: Ibid., part 3, ch. 2.

88 *Albert Einstein and his wife*: Marcia Bartusiak, "Einstein and Beyond," *National Geographic*, May 2005, available at http://science.nationalgeographic.com/science/space/universe/beyond-einstein.html.

89 *"All the books of Moses"*: John Redwood, *Reason, Ridicule, and Religion*, p. 119, and Roy Porter, *The Creation of the Modern World*, p. 130.

89 *"Is there anything more Absurd"*: Hunter, *Science and Society in Restoration England*, p. 175.

CHAPTER 15. A PLAY WITHOUT AN AUDIENCE

91fn *The moon gave the Greeks*: Jurgen Renn, ed., *Galileo in Context* (New York: Cambridge University Press, 2002), p. 198.

92fn *The stars will not look*: Albert Boime, "Van Gogh's *Starry Night*: A History of Matter and a Matter of History," *Arts Magazine*, December 1984, available at http://www.albertboime.com/Articles.cfm. Donald Olson, a Texas State University astronomer, has carried out similar work, notably a study of Edvard Munch's *The Scream*.

93 *"The falling body moved more jubilantly"*: Herbert Butterfield, *The Origins of Modern Science*, p. 6.

93 *"a book written in mathematical characters"*: The passage is from Galileo's

Assayer (1623), available at http://www.princeton.edu/~hos/h291/assayer. htm.

94fn *Galileo's intellectual offspring*: Richard Feynman, *The Character of Physical Law*, p. 58.

94 *"the actuality of a potentiality"*: Quoted in Joe Sachs, "Aristotle: Motion and Its Place in Nature," at http://www.iep.utm.edu/aris-mot/. The remark is quoted in slightly different form in Oded Balaban, "The Modern Misunderstanding of Aristotle's Theory of Motion," at http://tinyurl.com/y24yvwo.

94 *"If the ears, the tongue"*: Galileo, *The Assayer*.

95 *"communicate in the language"*: Charles Coulston Gillispie, *The Edge of Objectivity*, p. 43.

95 *"Do not all charms fly"*: John Keats, *Lamia*, part 2.

95 *"When I heard the learn'd astronomer"*: Walt Whitman, "When I Heard the Learn'd Astronomer."

96 *"Shut up and calculate"*: The remark is nearly always attributed to Feynman, it seems to have been coined by the physicist David Mermin. See David Mermin, "Could Feynman Have Said This?," *Physics Today*, May 2004, p.10, available at http://tinyurl.com/yz5qxhp.

96 *People do not "know a thing"*: Steven Nadler, "Doctrines of explanation in late scholasticism and in the mechanical philosophy," in Daniel Garber and Michael Ayers, eds., *The Cambridge History of Seventeenth-Century Philosophy* (New York: Cambridge University Press, 1998).

96 *"not a necessary part"*: Kline, *Mathematics: The Loss of Certainty*, p. 47, quoting Galileo, *Two New Sciences*.

CHAPTER 16. ALL IN PIECES

97 *"It is not only the heavens"*: Richard Westfall, "Newton and the Scientific Revolution," in Stayer, ed., *Newton's Dream*, p. 10.

98 *seventeenth-century Italy feared science*: Some recent scholars have argued that this notion is out of date. "The older Italian historiography tended to present late seventeenth-century science as sucked back in time by the black hole of Galileo's trial," writes Mario Biagioli, but "recent work has shown that such a simple explanation will not do." See Roy Porter and Mikulas Teich, eds., *The Scientific Revolution in National Context* (New York: Cambridge University Press, 1992), p. 12.

98 *"for at the slightest jar"*: Thomas Kuhn, *The Copernican Revolution*, p. 190, quoting Jean Bodin.

99 *"Worst of all"*: Ibid., p. 193.

99 *"Sense pleads for Ptolemy"*: Kline, *Mathematics in Western Culture*, p. 117.

101 *With no other rationale*: Richard Westfall, "Newton and the Scientific Revolution," pp. 6–7.

101 *"If the moon, the planets"*: Arthur Koestler, *The Sleepwalkers*, p. 498.

102 *"The Sun is lost"*: John Donne, "An Anatomy of the World."

CHAPTER 17. NEVER SEEN UNTIL THIS MOMENT

105 *"on or about December 1910"*: Virginia Woolf, "Character in Fiction." Woolf had in mind how writers like James Joyce portrayed their characters' inner lives.

105 *"The Mathematical Professor at Padua"*: Nicolson, "The 'New Astronomy' and English Imagination," p. 35.

106 *He had known "all the stars"*: Kitty Ferguson, *Tycho and Kepler*, p. 46.

106 *"the greatest wonder"*: Ibid., p. 47.

107 *a standing-room-only crowd*: Nicolson, "The Telescope and Imagination," p. 8.

107 *On the morning of September 3, 1609*: New-York Historical Society Collections, 2nd ser. (1841), vol. 1, pp. 71–74. This is from an excerpt online at http://historymatters.gmu.edu/d/5829.

108 *The breakthrough that made the telescope*: Albert Van Helden, ed., in his "Introduction" to *Sidereal Nuncius* (The Sidereal Messenger), by Galileo Galilei (Chicago: University of Chicago Press, 1989), pp. 2–3.

108 *They took turns peering*: Ibid., p. 6.

108 *"Many of the nobles"*: Nicolson, "The Telescope and Imagination," p. 12.

108 *"to discover at a much greater distance"*: Van Helden, "Introduction," p. 7.

109 *It revealed true features*: Shapin, *The Scientific Revolution*, p. 72. I owe to Shapin these observations about the telescope having had to prove its trustworthiness. Shapin also cites a variety of other factors that made the telescope hard to use and hard to evaluate.

109 *Galileo continued to improve*: Van Helden, "Introduction," p. 9.

109 *"absolute novelty"*: The quotes from Galileo in this paragraph and in the next several sentences come from Nicolson, "The Telescope and Imagination," pp. 14–15.

111 *Why could not the Earth itself?*: Kuhn, *The Copernican Revolution*, p. 222.

111 *What could be "more splendid"*: Lovejoy, *The Great Chain of Being*, p. 126.

111 *"When the heavens were a little"*: Ibid., p. 133.

112 *"The eternal silence of these infinite spaces"*: Ibid., p. 127.

112 *Man occupied "the filth and mire"*: Ibid., p. 102. E. M. W. Tillyard, in *The Elizabethan World Picture*, writes that "the earth in the Ptolemaic system was the cesspool of the universe" (p. 39).

113 *Galileo's adversary Cardinal Bellarmine*: Karen Armstrong, *A History of God*, p. 290.

CHAPTER 18. FLIES AS BIG AS A LAMB

114 *He put his own saliva*: On September 17, 1683, Leeuwenhoek described his teeth-cleaning routine and the "animalcules" he found in his mouth. Excerpts from that letter, and much other material related to Leeuwenhoek's discoveries, can be found at http://ucmp.berkely.edu/history/leeuwen hoek.htm.

115 *"exceedingly small animals"*: Marjorie Nicolson, "The Microscope and English Imagination," p. 167.

115 *And he had witnesses*: Ibid., p. 167.

115 *the first person ever to see sperm cells*: *The Collected Letters of Antoni van Leeuwenhoek*, edited by a Committee of Dutch Scientists (Amsterdam: Swets & Zeitlinger, 1941), vol. 2, pp. 283–95. This letter was written in November 1677 to William Brouncker, president of the Royal Society.

116 *"His Majesty seeing the little animals"*: Clara Pinto-Correia, *The Ovary of Eve* (Chicago: University of Chicago Press, 1997), p. 69.

116 *"limbs with joints, veins in these limbs"*: Nicolson, "The Microscope and English Imagination," p. 210.

117 *"Were men and beast made"*: Michael White, *Isaac Newton: The Last Sorcerer*, p. 149, quoting a notebook entry of Newton's headed "Of God."

117 *"large Hollows and Roughnesses"*: Robert Hooke, *Micrographia*. See http://www.roberthooke.org.uk/rest5a.htm.

117 *"flies which look as big as a lamb"*: "Commentary on Galileo Galilei," in James Newman, ed., *The World of Mathematics*, vol. 2, p. 732fn.

118 *"a fine moss growing"*: Lisa Jardine, *The Curious Life of Robert Hooke*, p. 164.

118 *"one who walks about"*: Westfall, *Science and Religion in Seventeenth-Century England*, p. 27.

119 *"There may be as much curiosity"*: Shapin, *The Scientific Revolution*, p. 145.

CHAPTER 19. FROM EARTHWORMS TO ANGELS

120 *"Cubes, Rhombs, Pyramids"*: Nicolson, "The Microscope and English Imagination," p. 209, quoting Henry Baker, *Employment for the Microscope*. Baker wrote much later than Leeuwenhoek, in 1753, but everyone who has ever looked through a microscope has uttered some variant of Baker's remark.

121 *The central idea was that all the objects*: Tillyard, *The Elizabethan World Picture*, p. 26.

121 *"We must believe that"*: Ibid., p. 40.

122 *He strapped himself each day*: John Carey, "Pope's Fallibility," in *Original*

Copy: Selected Reviews and Journalism 1969–1986 (London: Faber & Faber, 1987), p. 109, and Harold Bloom, *Genius* (New York: Warner, 2002), p. 271.

123 *"The work of the creator"*: Lovejoy, *The Great Chain of Being*, p. 53.

123 *"worthy of an infinite CREATOR"*: Ibid., p. 133.

123 *"We must say that God"*: Ibid., p. 224.

124 *"If God had made use"*: Ibid., p. 179.

124 *"and the characters are triangles"*: Galileo, *The Assayer*.

124 *"Nature is pleased with simplicity"*: G. A. J. Rogers, "Newton and the Guaranteeing God," in Force and Popkin, eds., *Newton and Religion*, p. 232, quoting Newton's *Principia*.

124 *"It is impossible that God"*: Paolo Rossi, *Logic and the Art of Memory*, p. 193.

125 *"God always complies"*: Peter K. Machamer, *The Cambridge Companion to Galileo* (New York: Cambridge University Press, 1998), p. 193.

125 *"Nature does not make jumps"*: Robert Nisbet, *History of the Idea of Progress* (New York: Basic Books, 1980), p. 158.

125 *"If triangles had a god"*: Montesquieu, *Persian Letters*, no. 59.

125 *"Einstein was a man who"*: Jacob Bronowski, *The Ascent of Man*, p. 256.

CHAPTER 20. THE PARADE OF THE HORRIBLES

126 *"vast Multitude of different Sorts"*: John Ray, *The Wisdom of God Manifested in the Works of the Creation*, available at http://www.jri.org.uk/ray/wisdom/index.htm.

127 *"How extremely stupid"*: Leonard Huxley, *The Life and Letters of Thomas Henry Huxley* (New York: Appleton, 1916), vol. 1, p. 176.

127 *"It is natural to admit"*: André Maurois cites Voltaire's remark in his introduction to Voltaire's *Candide*, trans. Lowell Blair (New York: Bantam, 1959), p. 5.

128 *"Some kinds of beasts"*: Michael White, *Isaac Newton*, p. 149.

128 *The world contained wood*: Thomas, *Man and the Natural World*, p. 20.

128 *Even if someone had conceived*: Steve Jones, *Darwin's Ghost* (New York: Random House, 2000), p. 194.

128 *"a thought of God"*: David Dobbs, *Reef Madness: Charles Darwin, Alexander Agassiz, and the Meaning of Coral* (New York: Pantheon, 2005), p. 3.

CHAPTER 21. "SHUDDERING BEFORE THE BEAUTIFUL"

129 *"all things are numbers"*: Kline, *Mathematics: The Loss of Certainty*, p. 12.

129fn *As one of Pythagoras's followers*: Jamie James, *The Music of the Spheres* (New York: Springer, 1995), p. 35.

130 *"one of the truly momentous"*: Chandrasekhar, "Shakespeare, Newton, and Beethoven."

130 *St. Augustine explained*: Barrow, *Pi in the Sky*, p. 256.

131 *"the first scientific proof"*: Kline, *Mathematics: The Loss of Certainty*, p. 66.

132 *"You must have felt this, too"*: Chandrasekhar, "Shakespeare, Newton, and Beethoven."

133 *"shuddering before the beautiful"*: Ibid.

133 *"the years of searching"*: From a 1933 lecture by Einstein, "About the Origins of General Relativity," at Glasgow University. Matthew Trainer discusses Einstein's lecture in "About the Origins of the General Theory of Relativity: Einstein's Search for the Truth," *European Journal of Physics* 26, no. 6 (November 2005).

133 *"to watch the sunset"*: *The Autobiography of Bertrand Russell* (Boston: Little, Brown, 1967), p. 38.

133 *"Of all escapes from reality"*: Gian-Carlo Rota, *Indiscrete Thoughts*, p. 70.

134 *his head, impaled on a pike*: Ferguson, *Tycho and Kepler*, p. 344. My references to witches and Kepler's mother come from Ferguson and from Max Caspar, *Kepler*.

134 *"When the storm rages"*: Benson Bobrick, *The Fated Sky* (New York: Simon & Schuster, 2006), p. 70.

CHAPTER 22. PATTERNS MADE WITH IDEAS

135 *Mathematics had almost nothing*: For a brilliant account of the difference between math as a mathematician sees it and as the subject is taught in school, see Paul Lockhart, "A Mathematician's Lament," http://tinyurl.com/y89qbh9.

135 *"A mathematician, like a painter"*: G. H. Hardy, *A Mathematician's Apology*, p.13, available at http://math.boisestate.edu/~holmes/holmes/A%20Mathematician's%20Apology.pdf.

135 *"upon which Sir Isaac"*: Westfall, *Never at Rest*, p. 192.

136 *"A naturalist would scarce expect"*: Bronowski, *The Ascent of Man*, p. 227.

CHAPTER 23. GOD'S STRANGE CRYPTOGRAPHY

143 *If two dinosaurs*: Mario Livio, *Is God a Mathematician?*, p. 11, quoting Martin Gardner, *Are Universes Thicker than Blackberries?* (New York: Norton, 2004).

143 *"strange Cryptography"*: Nicolson, "The Telescope and Imagination," p. 6, quoting Sir Thomas Browne.

143 *Nature presented a greater challenge*: In an essay in 1930, Einstein wrote, "What a deep conviction of the rationality of the universe and what a yearn-

ing to understand Kepler and Newton must have had to enable them to spend years of solitary labor in disentangling the principles of celestial mechanics! Only one who has devoted his life to similar ends can have a vivid realization of what has inspired these men and given them the strength to remain true to their purpose in spite of countless failures. It is cosmic religious feeling that gives a man such strength." See Albert Einstein, "Religion and Science," *New York Times Magazine*, November 9, 1930.

144 God *"took delight to hide"*: Eamon, *Science and the Secrets of Nature*, p. 320.

CHAPTER 24: THE SECRET PLAN

145 *"In what manner does the countenance"*: Arthur Koestler, *The Sleepwalkers*, p. 279. Half a century after its publication, *The Sleepwalkers* remains the best and liveliest account of the birth of modern astronomy. I have drawn repeatedly on Koestler's superlative history.

145 *"I was born premature"*: Ibid., p. 231.

146 *"That man has in every way"*: Ibid., p. 236.

147 *The conjunction point after that*: My discussion of Jupiter and Saturn follows the account in Christopher M. Linton, *From Eudoxus to Einstein*, p. 170.

148 *"The delight that I took"*: Koestler, *The Sleepwalkers*, p. 247.

149 *"The triangle is the first"*: Ibid., p. 249.

CHAPTER 25. TEARS OF JOY

152 *"And now I pressed forward"*: Koestler, *The Sleepwalkers*, p. 250.

152 *"instead of twenty or one hundred"*: Ibid., p. 248.

153 *Euclid proved that there are exactly five*: One way to see that there can only be a limited number of Platonic solids is to focus on one vertex and imagine the faces that meet there. There must be at least three such faces, and the angles at each vertex must all be identical and must add up to less than 360 degrees. Meeting all those conditions at once is impossible unless each face is a triangle, square, or pentagon. (Each angle of a hexagon is 120 degrees, for instance, so three or more hexagons cannot meet at one vertex.)

153 *If you needed dice*: Marcus du Sautoy, *Symmetry* (New York: Harper, 2008), p. 5.

154 *He burst into tears*: Caspar, *Kepler*, p. 63.

154 *"Now I no longer regretted"*: Koestler, *The Sleepwalkers*, p. 251.

155 *"For a long time I wanted"*: Owen Gingerich, "Johannes Kepler and the New Astronomy," available at http://adsabs.harvard.edu/full/1972QJRAS..13..346G.

155 *He happily devoted*: Koestler, *The Sleepwalkers*, p. 269.

155 *"No one," he boasted*: Caspar, *Kepler*, p. 71.

155 *"too pretty not to be true"*: James Watson, *The Double Helix* (New York: Touchstone, 2001), p. 204.

156 *"Never in history"*: Gingerich, "Johannes Kepler and the New Astronomy," p. 350.

CHAPTER 26. WALRUS WITH A GOLDEN NOSE

157 *"Would that God deliver me"*: Rossi, *The Birth of Modern Science*, p. 70.

158 *"the heavenly motions are nothing but"*: Koestler, *The Sleepwalkers*, p. 392.

158fn *Not by the human ear*: Rattansi, "Newton and the Wisdom of the Ancients," p. 189.

158fn *The first person to refer*: Curtis Wilson, "Kepler's Laws, So-Called," *HAD News* (newsletter of the Historical Astronomy Division of the American Astronomical Society), no. 31, May 1994.

158 *"My brain gets tired"*: Giorgio de Santillana, *The Crime of Galileo*, p. 106fn.

159 *In his student days*: Ferguson, *Tycho and Kepler*, pp. 31–32.

160 *had cost a ton of gold*: Gingerich, "Johannes Kepler and the New Astronomy," p. 350.

160 *"any single instrument cost more"*: Koestler, *The Sleepwalkers*, p. 278.

160 *"I was in possession"*: Ibid., p. 345.

CHAPTER 27. CRACKING THE COSMIC SAFE

162 *Even armed with Tycho's*: By far the best account of the mathematical ins and outs is Koestler's *The Sleepwalkers*.

163 *But Tycho's data were twice*: Kuhn, *The Copernican Revolution*, pp. 211–12.

163 *"For us, who by divine kindness"*: Koestler, *The Sleepwalkers*, p. 322.

163 *"warfare" with the unyielding data*: Livio, *Is God a Mathematician?*, p. 249.

164 *Even Galileo, revolutionary though he was*: De Santillana, *The Crime of Galileo*, p. 106fn.

166 *"a cartload of dung"*: Koestler, *The Sleepwalkers*, p. 397.

167 *"On March 8 of this present year"*: Ibid., p. 394.

167 *"I have consummated the work"*: Ferguson, *Tycho and Kepler*, p. 340.

168 *He saw—somehow*: Joseph Mazur provides this example in *The Motion Paradox*, p. 91.

CHAPTER 28. THE VIEW FROM THE CROW'S NEST

169 *"I believe that if a hundred"*: Bertrand Russell, *The Scientific Outlook*, p. 34.

170 *"a way of bewitching"*: Quoted in de Santillana, *The Crime of Galileo*, p. 115.

170 *Galileo put the book away*: Ibid., pp. 106fn., 168.

171 *"He discourses often amid fifteen"*: Ibid., p. 112.

171 *"If reasoning were like hauling"*: Galileo, *The Assayer*.

175 *"Shut yourself up with some friend"*: Galileo, *Dialogue Concerning the Two Chief World Systems*. This discussion takes place on day two.

175 *"A company of chessmen"*: Locke, *Essay Concerning Human Understanding*, p. 98.

CHAPTER 29. SPUTNIK IN ORBIT, 1687

179 *"It has been observed that missiles"*: The passage is from Galileo's *Two New Sciences*, quoted in David Goodstein and Judith Goodstein, *Feynman's Lost Lecture*, p. 38.

181 *Newton pictured it all*: Newton drew the diagram in the 1680s, but it was first published after his death, in *A Treatise of the System of the World*, a less mathematical treatment of the *Principia*. See John Roche, "Newton's *Principia*," in Fauvel et al., eds., *Let Newton Be!*, p. 58.

CHAPTER 30. HIDDEN IN PLAIN SIGHT

182 *"My aim is to show"*: Shapin, *The Scientific Revolution*, p. 33.

183fn *"Music," Leibniz wrote*: Kline, *Mathematics in Western Culture*, p. 287.

183 *"Galileo spent twenty years"*: Gillispie, *The Edge of Objectivity*, p. 42.

CHAPTER 31. TWO ROCKS AND A ROPE

187 *Unlike most legends*: Crease, *The Prism and the Pendulum*, p. 31.

188 *"In performing the experiment"*: Ibid., p. 32.

188 *When television shows a diver*: Barry Newman, "Now Diving: Sir Isaac Newton," *Wall Street Journal*, August 13, 2008.

CHAPTER 32. A FLY ON THE WALL

190 *"I sleep ten hours"*: Alfred Hooper, *Makers of Mathematics* (Vintage, 1948), p. 209.

192fn *One prominent historian calls it*: The historian was Salomon Bochner, in *The Role of Mathematics in the Rise of Science* (Princeton, NJ: Princeton University Press, 1966), p. 40. For more on the invention of the musical staff, see Alfred Crosby, *The Measure of Reality: Quantification and Western Society, 1250–1600* (New York: Cambridge University Press, 1997), pp. 142–44.

193 *"the greatest single step ever"*: Livio, *Is God a Mathematician?*, p. 86.

CHAPTER 33. "EUCLID ALONE HAS LOOKED ON BEAUTY BARE"

194 *known today as Cartesian coordinates*: Descartes' original presentation differed from the treatment that would become standard, but all the future changes were implicit in his version.

195 *"I do not enjoy speaking in praise"*: E. T. Bell, *The Development of Mathematics*, p. 139.

195 *"a notable advance in the history"*: Alfred North Whitehead, *Science and the Modern World*, p. 20. Scientists have now found that human infants and various nonhuman animals can count (they can distinguish between two M&Ms and three, for instance), but Whitehead's point was that it took a breakthrough to see that such concepts as "twoness" were worth identifying.

195 *"The point about zero"*: Newman, ed., *The World of Mathematics*, vol. 1, p. 442.

196 *Descartes wrestled to make sense*: Helena M. Pycior, *Symbols, Impossible Numbers, and Geometric Entanglements* (New York: Cambridge University Press, 2006), p. 82.

197 *Nor did it matter if the rock*: Eugene Wigner makes this point in his path-breaking essay "The Unreasonable Effectiveness of Mathematics in the Natural Sciences."

198 *If there were vacuums*: Butterfield, *The Origins of Modern Science*, p. 3.

198fn *The question of whether vacuums*: Russell Shorto, *Descartes' Bones* (New York: Doubleday, 2008), p. 25.

198 *"Only by imagining an impossible"*: A. Rupert Hall, *From Galileo to Newton*, p. 63. Hall cites the two passages from Galileo that I quote in his brilliant discussion of abstraction in science. See ibid., pp. 63–64. My comment about mathematics and abstraction in the final sentence of this chapter is also a paraphrase of Hall's argument on his p. 63.

CHAPTER 34. HERE BE MONSTERS!

202 *Albert of Saxony, a logician*: My discussion follows the one on pp. 52–55 of John Barrow's admirably lucid *The Infinite Book*.

CHAPTER 35. BARRICADED AGAINST THE BEAST

210 *For decades mathematicians had all tried*: Struik, *A Concise History of Mathematics*, pp. 101–9.

CHAPTER 37. ALL MEN ARE CREATED EQUAL

219 *Abraham Lincoln asked his listeners*: Lincoln made his remark on October 15, 1858 (and in at least one earlier speech) in his last debate with Stephen Douglas. The complete text is at http://www.bartleby.com/251/72.html.

222 *"The planet Mars comes close"*: Kline, *Mathematics in Western Culture*, p. 230.

223 *Perhaps infinitesimals were real but*: Carl Boyer, *The History of the Calculus and Its Conceptual Development*, p. 213.

223 *Leibniz tried to explain*: William Dunham, *The Calculus Gallery*, p. 24.

223 *"an enigma rather than"*: Leibniz's puzzled disciples were James and John Bernoulli, quoted in Kline, *Mathematics: The Loss of Certainty*, p. 137.

223 *"the ultimate ratio"*: Ibid., p. 135.

223 *"In mathematics the minutest"*: Ibid., p. 134.

223 calculus *is the Latin*: Donald Benson, *A Smoother Pebble: Mathematical Explorations*, p. 167.

224 *"For science it cannot be"*: George Berkeley, *The Analyst: or A Discourse Addressed to an Infidel Mathematician* (London, 1754), p. 34.

224 *Leibniz, boundlessly optimistic*: Dunham, *The Calculus Gallery*, p. 24, and Kline, *Mathematics: The Loss of Certainty*, p. 140.

224 *"Persist," d'Alembert advised*: Kline, *Mathematics: The Loss of Certainty*, p. 162.

CHAPTER 38. THE MIRACLE YEARS

226 *Calculus was in the air*: Most Newtonian scholars, including Newton's most careful biographer, Richard Westfall, and the preeminent expert on Newton's mathematical work, D. T. Whiteside, argue emphatically that Newton achieved his mathematical breakthroughs essentially on his own. For a contrary point of view, arguing that the influence of the Cambridge mathematician Isaac Barrow on Newton has been downplayed, see Mordechai Feingold's "Newton, Leibniz, and Barrow, Too: An Attempt at a Reinterpretation," *Isis* 84, no. 2 (June 1993), pp. 310–38.

226 *market called Stourbridge Fair*: Stourbridge Fair served as Bunyan's inspiration for Vanity Fair in *A Pilgrim's Progress*. See Edmund Venables, *Life of John Bunyan* (London: Walter Scott, 1888), p. 173.

226 *"The way to chastity"*: Gale Christianson, *In the Presence of the Creator: Isaac Newton and his Times*, p. 258.

227 *Newton "read it 'til"*: D. T. Whiteside, "Isaac Newton: Birth of a Mathematician," p. 58.

227 *"Read only the titles"*: Westfall, *Never at Rest*, p. 98.

228 *"The same year in May"*: Ibid., p. 143.

229 *"All this," he wrote*: Ibid.

229 *"If you haven't done"*: Author interview, in Edward Dolnick, "New Ideas and Young Minds," *Boston Globe*, April 23, 1984.

230 *"Age is, of course, a fever chill"*: Quoted in Dean Simonton, *Creativity in Science* (New York: Cambridge University Press, 2004), p. 68.

230 *"I know that when"*: Barrow, *Pi in the Sky*, p. 165.

230 *"Look at a composer"*: Author interview, in Dolnick, "New Ideas."

231 *"no old Men (excepting Dr. Wallis)"*: Westfall, *Never at Rest*, p. 139.

231 *From his earliest youth*: Gale Christianson, "Newton the Man—Again."

231 *"difficulty & ill success"*: Christianson, *In the Presence of the Creator*, p. 260.

232 *He took the Latin form*: Ackroyd, *Newton*, p. 39.

232 *"I will give thee the treasures"*: Christianson, *Isaac Newton*, p. 58. The verse is Isaiah 45:3.

232 *"The fact that he was unknown"*: Westfall, *Never at Rest*, p. 137.

232 *"In 1665, as he realized"*: Ibid., p. 138.

CHAPTER 39. ALL MYSTERY BANISHED

234 *In fact, though, Leibniz felt*: Since God was infinite, His creation was infinite as well, which meant that the process of finding new things to understand was never-ending. But this was a virtue, not a defect, because human happiness consisted in *constantly* finding new aspects of God's perfection to admire.

234 *"I don't know what I may seem"*: Westfall, *Never at Rest*, p. 863.

234 *"As a blind man has no idea"*: I. Bernard Cohen's translation of *Principia* (Berkeley: University of California Press, 1999), p. 428.

236 *"perhaps the most resolute champion"*: Ernst Cassirer, "Newton and Leibniz," p. 379.

CHAPTER 40. TALKING DOGS AND UNSUSPECTED POWERS

237 *"In the century of Kepler"*: C. H. Edwards, Jr., *The Historical Development of the Calculus*, p. 231.

237 *"an aptitude that was hard to find"*: Leibniz's letter can be found at www.leibniz-translations.com, a marvelous website run by the English philosopher Lloyd Strickland. See http://www.leibniz-translations.com/dog.htm, "Account of a Letter from Mr. Leibniz to the Abbé de St. Pierre, on a Talking Dog."

237 *"a museum of everything"*: Wiener, "Leibniz's Project."

238 *"I have so much that is new"*: Stewart, *The Courtier and the Heretic*, p. 256.

238 *"If controversies were to arise"*: Umberto Eco, *The Search for the Perfect Language*, p. 281. (See Chapter 14, "From Leibniz to the *Encyclopédie*.")

238 *Today a diligent team*: Author interview with Lawrence Carlin, philosophy department at the University of Wisconsin at Oshkosh, July 15, 2008.

238 *"Leibniz was one of the supreme"*: Russell, *History of Western Philosophy*, p. 581.

239fn *Unbeknownst to Leibniz*: Harriot's work on the telescope is discussed by Albert Van Helden in his "Introduction" to Galileo's *Sidereal Messenger*,

p. 9, and his mathematical work is discussed in the online journal *Plus*. See Anna Faherty, "Thomas Harriot, A Lost Pioneer," at http://plus.maths. org/issue50/features/faherty/.

239 *"A container shall be provided"*: George Dyson, *Darwin Among the Machines*, p. 37.

240 *Leibniz's knowledge of mathematics*: Joseph E. Hofmann, *Leibniz in Paris 1672–1676: His Growth to Mathematical Maturity*, p. 2.

240 *"I read [mathematics] almost"*: Dunham, *The Calculus Gallery*, p. 21.

241 *his correspondence alone consisted*: Stewart, *The Courtier and the Heretic*, p. 138.

242 *At an elegant dinner party*: A. Rupert Hall, *Philosophers at War*, p. 54.

242 *Or perhaps he decided*: The suggestions in this sentence and the next are from email correspondence with Simon Schaffer, a distinguished historian of science at Cambridge University, on September 27, 2009.

243 *"6accdae13eff7i319n4o4qrr4s8t12ux"*: Westfall, *Never at Rest*, p. 265.

243 *Leibniz made no mention*: Hall, *Philosophers at War*, p. 77.

CHAPTER 41. THE WORLD IN CLOSE-UP

252 *"the philosopher's stone that changed"*: Bell, *The Development of Mathematics*, p. 134.

CHAPTER 42. WHEN THE CABLE SNAPS

254 *We find good news*: My discussion here of position, speed, and acceleration draws heavily on Ian Stewart's elegantly written account in *Nature's Numbers*, pp. 50–52.

256 *"You can work out distances"*: Stewart, *Nature's Numbers*, p. 15.

257 *Proust's "little pieces of paper"*: Marcel Proust, *Swann's Way*, trans. Lydia Davis (New York: Viking, 2003), p. 51.

257 *Of all the ways to fire a cannon*: Paul Nahin, *When Least is Best*, p. 165. Nahin also discusses the physics of shooting a basketball.

258 *"as dawn compares to the bright"*: Dunham, *The Calculus Gallery*, p. 19, quoting James Gregory.

CHAPTER 43. THE BEST OF ALL POSSIBLE FEUDS

259 *"one of the chief geometers"*: Hall, *Philosophers at War*, p. 111. For any student of the Newton-Leibniz feud, Hall's book is the essential text.

259 *"I value my friends"*: Ibid., p. 112.

259 *"Taking Mathematicks from the beginning"*: Westfall, *Never at Rest*, p. 721.

260 *"the spectacle of the century"*: Boorstin, *The Discoverers*, p. 413.

261 *"round his brains such a thick crust"*: William Henry Wilkins, *The Love of*

an Uncrowned Queen: Sophia Dorothea, Consort of George I (New York: Duffield, 1906), p. 72.

262 *"When in good humour Queen Anne"*: Macaulay, *History of England*, vol. 5, p. 190.

262 *the king's only cultural interests*: Plumb, *The First Four Georges*, p. 41.

262 *The problems rose out of*: The best source for the tangled affairs of the Hanover court is www.gwleibniz.com, a website maintained by the University of Houston philosopher Gregory Brown. See http://www.gwleibniz.com/sophie_dorothea_celle/sophie_dorothea_celle.html.

264 *"I dare say," Leibniz wrote*: Gregory Brown, "Personal, Political, and Philosophical Dimensions of the Leibniz-Caroline Correspondence," p. 271.

264 *"The king has joked"*: Ibid., p. 292.

265 *"perhaps the most famous"*: Ibid., p. 262.

265 *The princess scolded her ex-tutor*: Ibid., p. 282.

265 *"the great men of our century"*: Quoted at http://www.gwleibniz.com/caroline_ansbach/caroline.html.

265 *"What difference does it make"*: Brown, "Leibniz-Caroline Correspondence," p. 282.

CHAPTER 44. BATTLE'S END

266 *"attempted to rob me"*: Cited in Robert Merton's classic essay "Priorities in Scientific Discovery: A Chapter in the Sociology of Science," p. 635. Galileo's charge comes at the very beginning of *The Assayer*.

266 *"I certainly should be vexed"*: Merton, "Priorities in Scientific Discovery," p. 648.

267 *"Almost no one is capable"*: Alfred Adler, "Mathematics and Creativity," *New Yorker*, February 19, 1972.

269 *"I throw myself"*: Westfall, *Never at Rest*, p. 724.

269 *"numerous and skilful"*: Ibid., p. 725.

270 *"Mr. Leibniz cannot be"*: The entire review is reprinted as an appendix to Hall's *Philosophers at War*. The quoted passage appears on p. 298.

270 *It, too, was written*: Charles C. Gillispie, "Isaac Newton," in *Dictionary of Scientific Biography* (New York: Scribner's, 1970–80), vol. 10.

270 *"broke Leibniz' heart"*: William Whiston, *Historical Memoirs of the Life and Writings of Dr. Samuel Clarke* (London, 1748), p. 132.

CHAPTER 45. THE APPLE AND THE MOON

271 *"So few went to hear Him"*: Westfall, *Never at Rest*, p. 209.

272 *"In the year 1666"*: Ibid., p. 154.

272 *The story, which is the one thing*: Westfall discusses the evidence pro and

con in *Never at Rest*, pp. 154–55, and is more inclined than many to give the story some credence.

272 *Despite his craving:* Simon Schaffer, "Somewhat Divine," *London Review of Books*, November 16, 2000, reviewing I. Bernard Cohen's translation of Newton's *Principia*.

272 *Historians who have scrutinized:* See Cohen's "Introduction" to his translation of the *Principia*, p. 15, and Schaffer, "Somewhat Divine."

273 *"I began to think":* Westfall, *Never at Rest*, p. 143.

275 *By combining Kepler's third law:* I. Bernard Cohen, "Newton's Third Law and Universal Gravity," p. 572.

277 *"compared the force required":* Westfall, *Never at Rest*, p. 143.

CHAPTER 46. A VISIT TO CAMBRIDGE

279 *In crowded rooms thick:* Steven Shapin, "At the Amsterdam," *London Review of Books*, April 20, 2006, reviewing *The Social Life of Coffee* by Brian Cowan. See also Mark Girouard, *Cities and People* (New Haven, CT: Yale University Press, 1985), p. 207.

279 *Wren, still more skilled, confessed:* Merton, "Priorities in Scientific Discovery," p. 636.

279 *"Mr. Hook said that he had it":* Roche, "Newton's *Principia*," in Fauvel et al., eds., *Let Newton Be!*, p. 58.

280 *taverns with Peter the Great:* Manuel, *A Portrait of Isaac Newton*, p. 318.

280 *he would invent a diving bell:* Alan Cook, *Edmond Halley: Charting the Heavens and the Seas* (New York: Oxford University Press, 1998), pp. 11, 140–41, 281.

280 *"Sir Isaac replied immediately":* Westfall, *Never at Rest*, p. 403.

CHAPTER 47. NEWTON BEARS DOWN

282fn *The statement if a planet:* Bruce Pourciau, "Reading the Master: Newton and the Birth of Celestial Mechanics," and Curtis Wilson, "Newton's Orbit Problem."

283 *Albert Einstein kept a picture:* Dudley Herschbach, "Einstein as a Student," available at http://tinyurl.com/yjptcq8.

283 *"Nature to him":* This was from Einstein's foreword to a new edition of Newton's *Opticks*, published in 1931.

283 *"Now I am upon":* Westfall, *Never at Rest*, p. 405.

283 *"I never knew him take":* Ibid., p. 192.

284 *"When he has sometimes taken":* Ibid., p. 406.

284 *If everything attracted everything:* Kuhn, *The Copernican Revolution*, p. 258.

285 *"To do this business right":* Westfall, *Never at Rest*, p. 409.

285 *"That all these problems"*: Chandrasekhar, "Shakespeare, Newton, and Beethoven."

287 *"swallowed up and lost"*: Westfall, *Never at Rest*, p. 456.

CHAPTER 48. TROUBLE WITH MR. HOOKE

288 *"a nice man to deal with"*: Henry Richard Fox Bourne, *The Life of John Locke*, vol. 2 (New York: Harper Brothers, 1876), p. 514.

289 *"There is one thing more"*: Westfall, *Never at Rest*, p. 446.

289 *"Mr Hook seems to expect"*: Manuel, *A Portrait of Isaac Newton*, p. 154.

289 *"He has done nothing"*: Ibid., p. 155.

289 *"Philosophy [i.e., science] is such"*: Ibid., p. 155.

290 *He never replied to Hooke's letter*: Westfall, *Never at Rest*, pp. 387–88.

290 *Newton had designed a telescope*: Ibid., p. 233.

291 *"poore & solitary endeavours"*: Ibid., p. 237.

291 *"the oddest, if not the most considerable"*: Ibid., p. 237.

292 *"Now is not this very fine"*: Ibid., p. 448.

292 *Hooke stalked out of the room*: Manuel, *A Portrait of Isaac Newton*, p. 159.

292 *Even twenty years after*: Ibid., p. 137.

292 *In the course of the move*: Christianson, *Isaac Newton*, p. 106.

CHAPTER 49. THE SYSTEM OF THE WORLD

293 *"I must now again beg you"*: Westfall, *Never at Rest*, p. 450.

294 *If the universe had been governed by a different law*: Martin Rees, *Just Six Numbers*, p. 150. See also Schaffer, "Somewhat Divine."

295 *"Pick a flower on Earth"*: Dirac may have had in mind a line from Francis Thompson's poem "The Mistress of Vision," where Thompson writes that "thou canst not stir a flower without troubling of a star." The same thought had moved Edgar Allan Poe to shake his head at the audacity of Newton's theory of cosmic connectedness. "If I venture to displace, by even the billionth part of an inch, the microscopical speck of dust which lies now upon the point of my finger," Poe marveled in his essay "Eureka," " . . . I have done a deed which shakes the Moon in her path, which causes the Sun to be no longer the Sun, and which alters forever the destiny of the multitudinous myriads of stars that roll and glow in the majestic presence of their Creator."

295 *The Principia made its first appearance*: Samuel Pepys was president of the Royal Society in 1687, and his name appears on the title page just below Newton's.

296 *"Nearer the gods no mortal may approach"*: Westfall, *Never at Rest*, p. 437.

296 *the French astronomer Lagrange declared*: Morris Kline, *Mathematics in Western Culture*, p. 209.

296 *It began paying Halley*: Westfall, *Never at Rest*, p. 453.

CHAPTER 50. ONLY THREE PEOPLE

297 *"There goes the man that writt a book"*: Ibid., p. 468.

297 *The first print run was tiny*: Ackroyd, *Newton*, p. 89.

297 *"It is doubtful," wrote the historian*: Gillispie, *The Edge of Objectivity*, p. 140.

297 *Perhaps half a dozen scientists*: Hall, *Philosophers at War*, p. 52.

298 *"A Book for 12 Wise Men"*: "Lights All Askew in the Heavens," *New York Times*, November 9, 1919, p. 17. See http://tinyurl.com/ygpam73.

298 *"I'm trying to think who"*: Stephen Hawking, *A Brief History of Time* (New York: Bantam, 1998), p. 85.

298 *But he rarely mentions calculus*: I. Bernard Cohen discusses in detail Newton's use of calculus in the "Introduction" to his translation of the *Principia*, pp. 122–27.

298 *"Newton's geometry seems to shriek"*: Roche, "Newton's *Principia*," in Fauvel et al., eds., *Let Newton Be!*, p. 50.

299 *"By the help of the new Analysis"*: Westfall, *Never at Rest*, p. 424.

299 *"There is no letter"*: Cohen, "Introduction," p. 123.

300 *"As we read the* Principia*"*: Chandrasekhar, "Shakespeare, Newton, and Beethoven."

CHAPTER 51. JUST CRAZY ENOUGH

301 *Molière long ago made fun*: Thomas Kuhn famously cited Molière in *The Structure of Scientific Revolutions*, p. 104.

302 *"We are all agreed that your theory is crazy"*: Bohr made the remark to Wolfgang Pauli and added, "My own feeling is that it is not crazy enough." Dael Wolfle, ed., *Symposium on Basic Research* (Washington, DC: American Association for the Advancement of Science, 1959), p. 66.

302fn *In time, this bewilderment*: J. J. MacIntosh, "Locke and Boyle on Miracles and God's Existence," p. 196.

303 *"He claims that a body attracts"*: Brown, "Leibniz-Caroline Correspondence," p. 273.

303 *"Mysterious though it was"*: John Henry, "Pray do not Ascribe that Notion to me: God and Newton's Gravity," in Force and Popkin, eds., *The Books of Nature and Scripture*, p. 141.

303 *"even if an angel"*: Brown, "Leibniz-Caroline Correspondence," p. 291.

304 *If the sun suddenly exploded*: Brian Greene, *The Elegant Universe* (New York: Norton, 1999), p. 56.

305 *"so great an absurdity"*: Westfall, *Never at Rest*, p. 505.

305 *"To tell us that every Species"*: From the end of *Opticks*, quoted in Kuhn, *The Copernican Revolution*, p. 259.

306 *"as if it were a Crime"*: Westfall, *Never at Rest*, p. 779.

306 *"Ye cause of gravity"*: Ibid., p. 505.

306 *"I have not been able to discover"*: Cohen's translation of the *Principia*, p. 428.

CHAPTER 52. IN SEARCH OF GOD

307fn *The debate over whether*: "Hacked E-Mail Is New Fodder for Climate Change Dispute," *New York Times*, November 21, 2009.

308 *"He is eternal and infinite"*: Cohen's translation of the *Principia*, p. 427.

309 *"Scientists, like whoring Jerusalem"*: Dennis Todd, "Laputa, the Whore of Babylon, and the Idols of Science," *Studies in Philology* 75, no. 1 (Winter 1978), p. 113.

310 *"they may do any thing"*: Quoted in a brilliant, far-ranging essay by Steven Shapin, "Of Gods and Kings: Natural Philosophy and Politics in the Leibniz-Clarke Disputes," p. 211.

311 *If you stopped to think about it, wrote Whiston*: Todd, "Laputa, the Whore of Babylon, and the Idols of Science," p. 108.

311 *"Sir Isaac Newton, and his followers"*: Westfall, *Never at Rest*, p. 778.

311 *"If God does not concern himself"*: Shapin, "Of Gods and Kings," p. 193.

312 *"If a king had a kingdom"*: I owe this observation about Leibniz and politics to Martin Tamny, "Newton, Creation, and Perception," p. 54.

312 *Newton emphasized God's will*: Shapin, "Of Gods and Kings," p. 194.

CONCLUSION

315 *a French mathematician named Urbain Le Verrier*: Kline, *Mathematics: The Loss of Certainty*, pp. 62–63, and Kline, *Mathematics in Western Culture*, p. 210.

316 *Benjamin Franklin sat deep in thought*: Bernard Bailyn, *To Begin the World Anew* (New York: Vintage, 2004), pp. 71–73.

316 *"The Constitution of the United States"*: I. Bernard Cohen, *Science and the Founding Fathers*, p. 90.

317 *"I had no need of that hypothesis"*: Kline, *Mathematics in Western Culture*, p. 210.

317 *"Mr. Leibniz is dead"*: Westfall, *Never at Rest*, p. 779.

317 *"Nothing could give me a greater"*: Brown, "Leibniz-Caroline Correspondence," p. 285.

318 *"You would have thought it was a felon"*: Stewart, *The Courtier and the Heretic*, p. 306.

318 *"The more I got to know Leibniz"*: Ibid., p. 117, quoting Eike Hirsch.

319 *"stone dolls"*: Milo Keynes discusses Newton's views on art and literature in "The Personality of Isaac Newton," pp. 26–27.

319 *"If we evolved a race of Isaac Newtons"*: from an interview with Huxley in J. W. N. Sullivan, *Contemporary Mind* (London: Toulmin, 1934), p. 143.

319 *"The more I learned"*: I interviewed Westfall in connection with an article marking the three hundredth anniversary of the *Principia*. See Edward Dolnick, "Sir Isaac Newton," *Boston Globe*, July 27, 1987. Westfall used the same "wholly other" phrase in the preface to *Never at Rest*, p. x, where he discussed Newton's uniqueness in a bit more detail.

320 *"He cried out with admiration"*: Westfall, *Never at Rest*, p. 473.

320 *His fellow professors did not know*: Ibid., p. 194.

BIBLIOGRAPHY

Ackroyd, Peter. *Newton*. New York: Doubleday, 2006.

Adler, Alfred. "Mathematics and Creativity." *New Yorker*, February 19, 1972.

Armstrong, Karen. *A History of God: The 4000-Year Quest of Judaism, Christianity, and Islam*. New York: Ballantine, 1993.

Ashenburg, Katherine. *The Dirt on Clean*. New York: North Point, 1997.

Atkins, Peter. *Galileo's Finger: The Ten Great Ideas of Science*. New York: Oxford University Press, 2003.

Barrow, John. *The Infinite Book: A Short Guide to the Boundless, Timeless, and Endless*. New York: Vintage, 2006.

Barzun, Jacques. *From Dawn to Decadence: 500 Years of Western Cultural Life*. New York: HarperCollins, 2000.

Beer, Anna. *Milton: Poet, Pamphleteer, and Patriot*. London: Bloomsbury, 2009.

Bell, E. T. *The Development of Mathematics*. New York: McGraw-Hill, 1945.

Benson, Donald. *A Smoother Pebble: Mathematical Explorations*. New York: Oxford University Press, 2003.

Berlin, Isaiah. *The Age of Enlightenment*. Boston: Houghton Mifflin, 1956.

Berlinski, David. *Infinite Ascent: A Short History of Mathematics*. New York: Modern Library, 2008.

Blackburn, Simon. *Think*. New York: Oxford University Press, 1999.

Bochner, Salomon. *The Role of Mathematics in the Rise of Science*. Princeton, NJ: Princeton University Press, 1979.

Bondi, Hermann. *Relativity and Common Sense: A New Approach to Einstein*. New York: Dover, 1962.

Boorstin, Daniel. *The Discoverers*. New York: Random House, 1983.

Boyer, Carl. *The History of the Calculus and Its Conceptual Development*. New York: Dover, 1949.

Bronowski, Jacob. *The Ascent of Man*. Boston: Little, Brown, 1973.

Brooke, John. "The God of Isaac Newton." In Fauvel et al., eds., *Let Newton Be!*

Brown, Gregory. "Personal, Political, and Philosophical Dimensions of the

Leibniz-Caroline Correspondence." in Paul Lodge, ed., *Leibniz and His Correspondents*. New York: Cambridge University Press, 2004.

Burtt, E. A. *The Metaphysical Foundations of Modern Science*. New York: Doubleday, 1954.

Butterfield, Herbert. *The Origins of Modern Science*. New York: Macmillan, 1953.

Cantor, Norman. *In the Wake of the Plague*. New York: Simon & Schuster, 2001.

Carey, John. *John Donne: Life, Mind, and Art*. London: Faber & Faber, 1981.

Caspar, Max. *Kepler*. New York: Dover, 1993.

Cassirer, Ernst. "Newton and Leibniz." *Philosophical Review* 52, no. 4 (July 1943), pp. 366–91.

Chandrasekhar, S. "Shakespeare, Newton, and Beethoven, or Patterns in Creativity." http://www.sawf.org/newedit/edit02192001/musicarts.asp.

Chapman, Allan. *England's Leonardo: Robert Hooke and the Seventeenth-Century Scientific Revolution*. New York: Taylor & Francis, 2004.

Christianson, Gale. *In the Presence of the Creator: Isaac Newton and His Times*. New York: Free Press, 1984.

———. *Isaac Newton*. New York: Oxford University Press, 2005.

———. "Newton the Man—Again." In Paul Scheurer and G. Debrock, eds., *Newton's Scientific and Philosophical Legacy*. New York: Springer, 1988.

Cockayne, Emily. *Hubbub: Filth, Noise and Stench in England*. New Haven, CT: Yale University Press, 2007.

Cohen, I. Bernard. "Newton's Third Law and Universal Gravity." *Journal of the History of Ideas* 48, no. 4 (October–December 1987), pp. 571–93.

———. *Revolution in Science*. Cambridge, MA: Harvard University Press, 1985.

———. *Science and the Founding Fathers: Science in the Political Thought of Thomas Jefferson, Benjamin Franklin, John Adams, and James Madison*. New York: Norton, 1997.

Cook, Alan. "Halley and Newton's *Principia*." *Notes and Records of the Royal Society of London* 45, no. 2 (July 1991), pp. 129–38.

Crease, Robert. *The Prism and the Pendulum: The Ten Most Beautiful Experiments in Science*. New York: Random House, 2003.

Dantzig, Tobias. *Number: The Language of Science*. New York: Macmillan, 1954.

Daston, Lorraine, and Katharine Park. *Wonders and the Order of Nature, 1150–1750*. New York: Zone, 2001.

Davis, Martin. *The Universal Computer*. New York: Norton, 2000.

Davis, Philip, and Reuben Hersh. *The Mathematical Experience*. Boston: Birkhauser, 1981.

De Santillana, Giorgio. *The Crime of Galileo*. Chicago: University of Chicago Press, 1955.

Drake, Stillman. "The Role of Music in Galileo's Experiments." *Scientific American,* June 1975.

Dunham, William. *The Calculus Gallery: Masterpieces from Newton to Lebesgue.* Princeton, NJ: Princeton University Press, 2005.

———. *Journey Through Genius: The Great Theorems of Mathematics.* New York: Penguin, 1990.

Dyson, George. *Darwin Among the Machines.* Reading, MA: Perseus, 1997.

Eamon, William. *Science and the Secrets of Nature: Books of Secrets in Medieval and Early Modern Culture.* Princeton, NJ: Princeton University Press, 1994.

Eco, Umberto. *The Search for the Perfect Language.* Waukegan, IL: Fontana, 1997.

Edwards, C. H., Jr. *The Historical Development of the Calculus.* New York: Springer-Verlag, 1979.

Einstein, Albert. "Religion and Science." *New York Times Magazine,* November 9, 1930, pp. 1–4. Reprinted in Einstein's *Ideas and Opinions* and *The World As I See It.*

Eves, Howard. *An Introduction to the History of Mathematics.* New York: Holt, Rinehart & Winston, 1964.

Fauvel, John, Raymond Flood, Michael Shortland, and Robin Wilson, eds. *Let Newton Be!* New York: Oxford University Press, 1988.

Feingold, Mordechai. *The Newtonian Moment: Isaac Newton and the Making of Modern Culture.* New York: Oxford University Press, 2004.

Ferguson, Kitty. *Tycho and Kepler: The Unlikely Partnership That Forever Changed Our Understanding of the Heavens.* New York: Walker, 2002.

Feynman, Richard. *The Character of Physical Law.* Cambridge, MA: MIT Press, 1967.

Force, James. "Newton, the 'Ancients,' and the 'Moderns.'" In Force and Popkin, eds., *Newton and Religion.*

Force, James, and Richard Popkin, eds. *The Books of Nature and Scripture: Recent Essays on Natural Philosophy, Theology, and Biblical Criticism in the Netherlands of Spinoza's Time and the British Isles of Newton's Time.* Dordrecht, Netherlands: Kluwer, 1994.

———, eds. *Newton and Religion: Context, Nature, and Influence.* Dordrecht, Netherlands: Kluwer, 1999.

Fraser, Antonia. *Royal Charles: Charles II and the Restoration.* New York: Delta, 1980.

Gillispie, Charles Coulston. *The Edge of Objectivity.* Princeton, NJ: Princeton University Press, 1960.

Gingerich, Owen. "Johannes Kepler and the New Astronomy." *Quarterly Journal of the Royal Astronomical Society* 13 (1972), pp. 346–73.

Golinski, Jan. "The Secret Life of an Alchemist." In Fauvel et al., eds., *Let Newton Be!*

Goodstein, David, and Judith Goodstein. *Feynman's Lost Lecture: The Motion of Planets Around the Sun.* New York: Norton, 1996.

Hadamard, Jacques. *The Psychology of Invention in the Mathematical Field.* New York: Dover, 1954.

Hahn, Alexander. *Basic Calculus: From Archimedes to Newton to Its Role in Science.* New York: Springer, 1998.

Hall, A. Rupert. *From Galileo to Newton.* New York: Dover, 1981.

———. *Philosophers at War: The Quarrel Between Newton and Leibniz.* New York: Cambridge University Press, 1980.

Hanson, Neil. *The Great Fire of London.* Hoboken, NJ: Wiley, 2001.

Hardy, G. H. *A Mathematician's Apology.* In Newman, ed., *The World of Mathematics,* vol. 4, available at http://math.boisestate.edu/~holmes/holmes/A%20 Mathematician's%20Apology.pdf.

Healy, Margaret. "Defoe's *Journal* and the English Plague Writing Tradition." *Literature and Medicine* 22, no. 1 (2003), pp. 25–55.

Henry, John. "Newton, Matter, and Magic." In Fauvel et al., eds., *Let Newton Be!*

———. "Occult Qualities and the Experimental Philosophy: Active Principles in Pre-Newtonian Matter Theory." *History of Science* 24 (1986), pp. 335–81.

———. "Pray do not Ascribe that Notion to me: God and Newton's Gravity." In Force and Popkin, eds., *The Books of Nature and Scripture.*

Hofmann, Joseph. *Leibniz in Paris 1672–1676: His Growth to Mathematical Maturity.* New York: Cambridge University Press, 2008.

Hollis, Leo. *London Rising: The Men Who Made Modern London.* New York: Walker, 2008.

Hoyle, Fred. *Astronomy and Cosmology: A Modern Course.* San Francisco: W. H. Freeman, 1975.

Hunter, Michael, ed. *Robert Boyle Reconsidered.* New York: Cambridge University Press, 1994.

———. *Science and Society in Restoration England.* New York: Cambridge University Press, 1981.

Iliffe, Rob. "Butter for Parsnips: Authorship, Audience, and the Incomprehensibility of the *Principia.*" In Mario Biagioli and Peter Galison, eds., *Scientific Authorship: Credit and Intellectual Property in Science.* New York: Routledge, 2003.

———. "'In the Warehouse': Privacy, Property and Priority in the Early Royal Society." *History of Science* 30 (1992), pp. 29–68.

———. "'Is he like other men?' The Meaning of the *Principia Mathematica,* and the author as idol." In Gerald Maclean, ed., *Culture and Society in the Stuart Revolution*. New York: Cambridge University Press, 1995.

Jardine, Lisa. *The Curious Life of Robert Hooke*. New York: HarperCollins, 2004.

———. *Ingenious Pursuits: Building the Scientific Revolution*. New York: Doubleday, 1999.

———. *On a Grander Scale: The Outstanding Life of Sir Christopher Wren*. New York: HarperCollins, 2002.

Johnson, George. *The Ten Most Beautiful Experiments*. New York: Knopf, 2008.

Jolley, Nicholas. *Leibniz*. New York: Routledge, 2005.

Kelly, John. *The Great Mortality: An Intimate History of the Black Death, the Most Devastating Plague of All Time*. New York: HarperCollins, 2005.

Keynes, John Maynard. "Newton, the Man." In Newman, ed., *The World of Mathematics*, vol. 1.

Keynes, Milo. "The Personality of Isaac Newton." *Notes and Records of the Royal Society of London* 49, no. 1 (January 1995), pp. 1–56.

Kline, Morris. *Mathematics: The Loss of Certainty*. New York: Oxford University Press, 1980.

———. *Mathematics in Western Culture*. New York: Oxford University Press, 1953.

Koestler, Arthur. *The Sleepwalkers*. New York: Grosset & Dunlap, 1970.

Koyré, Alexandre. *From the Closed World to the Infinite Universe*. Baltimore: Johns Hopkins University Press, 1957.

Kubrin, David. "Newton and the Cyclical Cosmos: Providence and the Mechanical Philosophy." *Journal of the History of Ideas* 28, no. 3 (July–September 1967), pp. 325–46.

Kuhn, Thomas. *The Copernican Revolution*. Cambridge, MA: Harvard University Press, 1957.

———. *The Structure of Scientific Revolutions*. Chicago: University of Chicago Press, 1996.

Linton, Christopher. *From Eudoxus to Einstein: A History of Mathematical Astronomy*. New York: Cambridge University Press, 2004.

Livio, Mario. *Is God a Mathematician?* New York: Simon & Schuster, 2009.

Lloyd, Claude. "Shadwell and the Virtuosi." *Proceedings of the Modern Language Association* 44, no. 2 (June 1929), pp. 472–94.

Lockhart, Paul. "A Mathematician's Lament." http://tinyurl.com/y89qbh9.

Lovejoy, Arthur. *The Great Chain of Being*. New York: Harper, 1960.

Lund, Roger. "Infectious Wit: Metaphor, Atheism, and the Plague in Eighteenth-Century London." *Literature and Medicine* 22, no. 1 (Spring 2003), pp. 45–64.

MacIntosh, J. J. "Locke and Boyle on Miracles and God's Existence." In Hunter, ed., *Robert Boyle Reconsidered*.

Manuel, Frank. *A Portrait of Isaac Newton*. Cambridge, MA: Belknap, 1968.
———. *The Changing of the Gods*. Hanover, NH: University Press of New England for Brown University Press, 1983.
Mazur, Joseph. *The Motion Paradox*. New York: Dutton, 2007.
McGuire, J. E., and P. M. Rattansi. "Newton and the 'Pipes of Pan.'" *Notes and Records of the Royal Society of London* 21, no. 2 (December 1966), pp. 108–43.
McNeill, William. *Plagues and Peoples*. New York: Doubleday, 1976.
Merton, Robert. *On the Shoulders of Giants*. New York: Free Press, 1965.
———. "Priorities in Scientific Discovery: A Chapter in the Sociology of Science." *American Sociological Review* 22, no. 6 (December 1957), pp. 635–59.
Merz, John Theodore. *Leibniz*. New York: Hacker, 1948.
Miller, Perry. "The End of the World." *William and Mary Quarterly*, 3rd ser., vol. 8, no. 2 (April 1951), pp. 172–91.
Moote, A. Lloyd, and Dorothy Moote. *The Great Plague*. Baltimore: Johns Hopkins University Press, 2004.
Nadler, Steven. *The Best of All Possible Worlds: A Story of Philosophers, God, and Evil*. New York: Farrar, Straus & Giroux, 2008.
Nahin, Paul. *When Least Is Best: How Mathematicians Discovered Many Clever Ways to Make Things as Small (or as Large) as Possible*. Princeton, NJ: Princeton University Press, 2007.
Newman, James, ed. *The World of Mathematics*. 4 vols. New York: Simon & Schuster, 1956.
Nicolson, Marjorie. "The Telescope and Imagination," "The 'New Astronomy' and English Imagination," "The Scientific Background of Swift's *Voyage to Laputa*" (with Nora Mohler), and "The Microscope and English Imagination." Separate essays reprinted in Marjorie Nicolson, *Science and Imagination*. Ithaca, NY: Cornell University Press, 1962.
Nicolson, Marjorie, and Nora Mohler. "Swift's 'Flying Island' in the *Voyage to Laputa*." *Annals of Science* 2, no. 4 (January 1937), pp. 405–30.
Pepys, Samuel. *The Diary of Samuel Pepys*. http://www.pepysdiary.com.
Pesic, Peter. "Secrets, Symbols, and Systems: Parallels Between Cryptanalysis and Algebra, 1580–1700." In Hunter, ed., *Robert Boyle Reconsidered*.
Picard, Liza. *Restoration London*. New York: Avon, 1997.
Porter, Roy. *The Creation of the Modern World*. New York: Norton, 2000.
———. *English Society in the Eighteenth Century*. New York: Penguin, 1982.
Pourciau, Bruce. "Reading the Master: Newton and the Birth of Celestial Mechanics." *American Mathematical Monthly* 104, no. 1 (January 1997), pp. 1–19.
Rattansi, Piyo. "Newton and the Wisdom of the Ancients." In Fauvel et al., eds., *Let Newton Be!*

Redwood, John. *Reason, Ridicule, and Religion: The Age of Enlightenment in England 1660–1750.* Cambridge, MA: Harvard University Press, 1976.

Rees, Martin. *Just Six Numbers: The Deep Forces That Shape the Universe.* New York: Basic Books, 2001.

Roche, John. "Newton's *Principia.*" In Fauvel et al., eds., *Let Newton Be!*

Rogers, G. A. J. "Newton and the Guaranteeing God." In Force and Popkin, eds., *Newton and Religion.*

Rossi, Paolo. *The Birth of Modern Science.* Malden, MA: Blackwell, 2001.

———. *Logic and the Art of Memory: The Quest for a Universal Language.* New York: Continuum, 2006.

Rota, Gian-Carlo. *Indiscrete Thoughts.* Boston: Birkhauser, 2008.

Russell, Bertrand. *The Scientific Outlook.* New York: Norton, 1962.

Schaffer, Simon. "Somewhat Divine." *London Review of Books,* November 16, 2000.

Seife, Charles. *Zero: The Biography of a Dangerous Idea.* New York: Penguin, 2000.

Shapin, Steven. "Of Gods and Kings: Natural Philosophy and Politics in the Leibniz-Clarke Disputes." *Isis* 72, no. 2 (June 1981), pp. 187–215.

———. "One Peculiar Nut." *London Review of Books,* January 23, 2003. (This is an essay on Descartes.)

———. "Rough Trade." *London Review of Books,* March 6, 2003. (This is an essay on Robert Hooke.)

———. "The House of Experiment in Seventeenth-Century England." *Isis* 79, no. 3 (September, 1988), pp. 373–404.

———. *A Social History of Truth: Civility and Science in Seventeenth Century England.* Chicago: University of Chicago Press, 1995.

———. *The Scientific Revolution.* Chicago: University of Chicago Press, 1996.

Smith, Virginia. *Clean: A History of Personal Hygiene and Purity.* New York: Oxford University Press, 2007.

Smolinski, Reiner. "The Logic of Millennial Thought: Sir Isaac Newton Among His Contemporaries." In Force and Popkin, eds., *Newton and Religion.*

Snobelen, Stephen. "Lust, Pride and Ambition: Isaac Newton and the Devil." In James Force and Sarah Hutton, eds., *Newton and Newtonianism: New Studies.* Dordrecht, Netherlands: Kluwer, 2004, pp. 155–81.

Stayer, Marcia Sweet, ed. *Newton's Dream.* Chicago: University of Chicago Press, 1988.

Stewart, Ian. *Nature's Numbers.* New York: Basic Books, 1995.

Stewart, Matthew. *The Courtier and the Heretic: Leibniz, Spinoza, and the Fate of God in the Modern World.* New York: Norton, 2006.

Stillwell, John. *Mathematics and Its History.* New York: Springer, 1989.

Stone, Lawrence. *The Family, Sex and Marriage in England 1500–1800.* New York: Penguin, 1979.

Struik, Dirk. *A Concise History of Mathematics.* New York: Dover, 1948.

Tamny, Martin. "Newton, Creation, and Perception." *Isis* 70, no. 1 (March 1979), pp. 48–58.

Thomas, Keith. *Man and the Natural World.* New York: Pantheon, 1983.

———. *Religion and the Decline of Magic.* New York: Scribner's, 1971.

Tillyard, E. M. W. *The Elizabethan World Picture.* New York: Vintage, 1961.

Tinniswood, Adrian. *His Invention So Fertile: A Life of Christopher Wren.* New York: Oxford University Press, 2001.

Tomalin, Claire. *Samuel Pepys.* New York: Knopf, 2002.

Weber, Eugen. *Apocalypses: Prophecies, Cults and Millennial Beliefs Through the Ages.* Cambridge, MA: Harvard University Press, 2000.

Weinberg, Steven. "Newton's Dream." In Stayer, ed., *Newton's Dream.*

Westfall, Richard S. *Never at Rest: A Biography of Isaac Newton.* New York: Cambridge University Press, 1980.

———. "Newton and the Scientific Revolution." In Stayer, ed., *Newton's Dream.*

———. *Science and Religion in Seventeenth-Century England.* Ann Arbor: University of Michigan Press, 1973.

———. "Short-Writing and the State of Newton's Conscience, 1662 (1)." *Notes and Records of the Royal Society of London* 18, no. 1 (June 1963), pp. 10–16.

White, Michael. *Isaac Newton: The Last Sorcerer.* Reading, MA: Perseus, 1997.

Whitehead, Alfred North. *Science and the Modern World.* New York: Free Press, 1925.

Whiteside, D. T. "Isaac Newton: Birth of a Mathematician." *Notes and Records of the Royal Society of London* 19, no. 1 (June 1964), pp. 53–62.

——— ed. *The Mathematical Papers of Isaac Newton.* Vol. 1, *1664–1666.* New York: Cambridge University Press, 1967.

Wiener, Philip. "Leibniz's Project of a Public Exhibition of Scientific Inventions." *Journal of the History of Ideas* 1, no. 2 (April 1940), pp. 232–40.

Wigner, Eugene. "The Unreasonable Effectiveness of Mathematics in the Natural Sciences." *Communications in Pure and Applied Mathematics* 13, no. 1 (February 1960), pp. 1–14.

Wilson, Curtis. "Newton's Orbit Problem: A Historian's Response." *College Mathematics Journal* 25, no. 3 (May 1994), pp. 193–200.

Wisan, Winifred. "Galileo and God's Creation." *Isis* 77, no. 3 (September 1986), pp. 473–86.

ILLUSTRATION CREDITS

Page 11 Public domain.

Page 12 Top: Portrait of Galileo Galilei (1564–1642), astronomer and physicist (drawing), by Ottavio Mario Leoni (c.1578–1630). Biblioteca Marucelliana, Florence, Italy/The Bridgeman Art Library.
Bottom: Gal. 48, fol. 28r, Firenze, Biblioteca Nazionale Centrale. Reproduced by kind permission of the Ministero per i Beni e le Attività Culturali, Italy/Biblioteca Nazionale Centrale, Firenze. This image cannot be reproduced in any form without the authorization of the Library, the owners of the copyright.

Page 13 Top: Pythagoras (c.580–500 BC) discovering the consonances of the octave, from "Theorica Musicae" by Franchino Gaffurio, first published in 1480, from 'Revue de l'Histoire du Theatre,' 1959 (engraving) (b/w photo) by French School (20th century). Bibliothèque des Arts Decoratifs, Paris, France/Archives Charmet/The Bridgeman Art Library.
Bottom: Pythagoras (c.580–500 BC), Greek philosopher and mathematician, Roman copy of Greek original (marble) by Pinacoteca Capitolina, Palazzo Conservatori, Rome, Italy/Index/The Bridgeman Art Library.

Page 14 Top: Public domain.
Bottom: Telescope belonging to Sir Isaac Newton (1642–1727), 1671 by English School.
Royal Society, London, UK/The Bridgeman Art Library.

Page 15 Top: Portrait of Edmond Halley, c.1687 (oil on canvas) by Thomas Murray (1663–1734).
Royal Society, London, UK/The Bridgeman Art Library.
Bottom: Wellcome Library, London.

Page 16 © Werner Forman/CORBIS.

INDEX